STUDENT SOLUTIONS MANUAL AND STUDY GUIDE for

Sellers, Crossno, Bright, Himmelmann:

An Introduction to Business Mathematics

Gene Sellers
Sacramento City College

SAUNDERS COLLEGE PUBLISHING
Philadelphia New York Chicago
San Francisco Montreal Toronto
London Sydney Tokyo Mexico City
Rio de Janeiro Madrid

Address orders to:
383 Madison Avenue
New York, NY 10017

Address editorial correspondence to:
West Washington Square
Philadelphia, PA 19105

Student Solutions Manual and Study Guide for AN INTRODUCTION TO BUSINESS MATHEMATICS

Volume ISBN 0-03-006128-8

78 018 98765432

CBS College Publishing
Saunders College Publishing
Holt, Rinehart and Winston
The Dryden Press

PREFACE

TO

SOLUTIONS MANUAL AND STUDY GUIDE

This Student Solutions Manual and Study Guide is a supplement to Business Mathematics by Sellers, Bright, Crossno and Himmelmann. The purpose of the Guide is to provide a student with additional assistance that is sometimes needed to cover the material given in a business mathematics course. This purpose is achieved by providing three sections for each Chapter in the text.

Part A Summary of Topics
Part B Selected Solutions
Part C Chapter (Number) Sample Test

In Part A, Summary of Topics, an outline is given of the major topics studied in the text. The outline is separated into sections that correspond to the ones used in the text. For example, the topics in Section 6-2, Discounts Based on Volume, are located in the study guide in the section similarly identified.

The format of the summary is to key on those topics or issues that are considered most important. Frequently words and phrases such as "notice", "remember", and "keep in mind" are used to focus attention on such issues. Occasionally some additional comments are given on some topic to provide further explanation, or cite additional examples, or simply repeat a definition using different words. All the formulas in the text are repeated in the study guide, and when appropriate, further explanation of a given formula is supplied. Thus, the summary is an attempt to extract from the text the key items and organize them in a compact form.

In Part B, Selected Solutions, about twenty-five percent of the problems from the Set A Exercise Sets are worked out in detail. Basically, problems 1, 5, 9, 13 and so on have these detailed solutions. Each complete solution illustrates the step-by-step process that can be used to work out a problem. Many of these solutions have marginal notes that identify a formula used, or an explanation as to what was done at a given point in the process of solving the problem. The twenty-five percent figure adequately covers a representative part of all of the different types of problems in each exercise set.

The purpose of this section is to give students some help on solving problems. It is hoped that students can then successfully work the other seventy-five percent of the problems in the text. In any mathematics course, learning to solve problems can (in most cases) only be accomplished by doing. Many students in mathematics become discouraged when they are unable to complete a homework assignment. Quite often the frustration associated with solving problems can be eliminated by one or two clues on what to do - or what not to do. These detailed solutions frequently provide the necessary assistance to overcome that frustration and anger.

The following sequence of steps is recommended for a proper use of the detailed solutions in the study guide.

Step 1. *Try to work the problem on your own.*
Step 2. *If you have difficulty figuring out what to do, check the example in the text that illustrates a solution for a similar problem.*
Step 3. *Now try again to work the problem in the exercise set.*
Step 4. *If you still cannot work the problem, check the detailed solution in the study guide to see what it was that prevented you from solving the problem.*
Step 5. *Work a similar problem that does not have a detailed solution in the study guide.*

Notice, since the Set B exercises may be used by instructors for graded homework assignments, there are no detailed solutions to any of these exercises.

Part C, a Chapter Sample Test, provides a student with an opportunity to take a test for practice. These tests are designed to cover the quantity of material that can reasonably be given for a one-hour examination. It is therefore important that the practice test be completed in the alloted time given for in-class tests. The answers for these tests are given in the end pages of the study guide.

TABLE OF CONTENTS

CHAPTER ONE

WHOLE NUMBERS AND COMMON FRACTIONS

PART A Summary of Topics

OBJECTIVES

1. *Write whole numbers in number and word forms.*
2. *Round off a whole number to an indicated place value.*
3. *Add whole numbers.*
4. *Subtract whole numbers.*
5. *Multiply whole numbers.*
6. *Divide whole numbers.*
7. *Learn the rule for order of operations.*
8. *Reduce common fractions.*
9. *Add common fractions.*
10. *Subtract common fractions.*
11. *Multiply common fractions.*
12. *Divide common fractions.*

SECTION 1-1 Reading and Writing Whole Numbers

1. The numbers 0, 1, 2, 3, 4, 5, 6, 7, 8 and 9 are called digits.

2. Place value means that a given digit can represent different quantities by virtue of its place in a number.

   5 means 5 units (or ones).
   50 means 5 groups of ten each, or simply 5 tens.
   5,000 means 5 groups of a thousand each, or simply 5 thousands.

3. The names of the first thirteen places are listed from right to left.

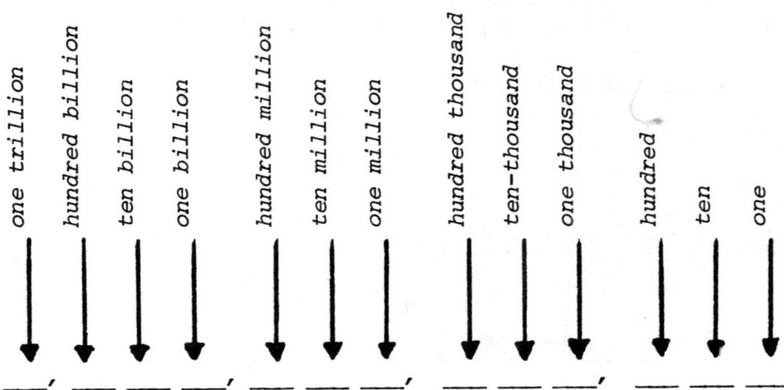

4. A number is usually written in number form by using the digits and place value.

5. A number can be written in word form by writing the number in words as it is read.

   A number on a check that is used for money is written both ways to insure that the amount of money represented by the check is correctly identified.

6. Two steps are needed to round (or round-off) a whole number:

   Step 1. Locate the digit (call it d) in the place to which the number is to be rounded.

SECTION 1-1 Reading and Writing Whole Numbers

If in the rounded number d is not changed, the number is said to be rounded down (that is, reduced). If in the rounded number d is replaced by d + 1, the number is said to be rounded up (that is, increased).

Step 2 Look only at the digit immediately to the right of d.

a. If this digit is 0, 1, 2, 3 or 4, then change all the digits to the right of d to zeros. (The number is rounded down).

b. If this digit is 5, 6, 7, 8 or 9, then add one to d and change all the digits to the right of d to zeros. (The number is consequently rounded up).

Whether a number is rounded up or rounded down, the new number is an approximation for the original number. Such rounding is frequently a matter of convenience, in that the rounded number has fewer nonzero digits and is easier to work with.

SECTION 1-2 Addition and Subtraction of Whole Numbers

1. When two numbers are added, the result is called the sum. The numbers added are called addends.

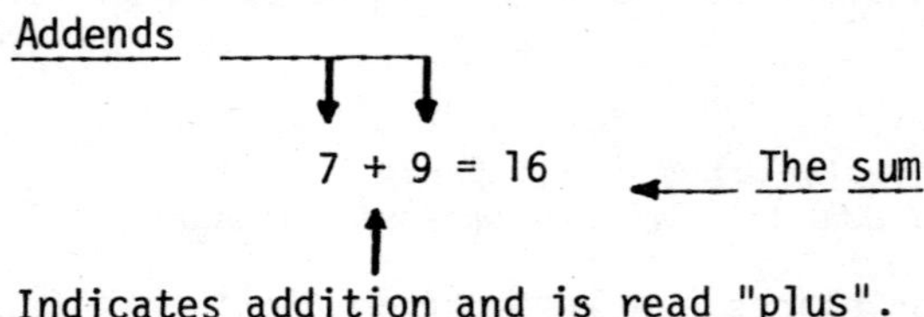

2. When two numbers are subtracted, the result is called the difference. The numbers in a subtraction problem are called the subtrahend and minuend.

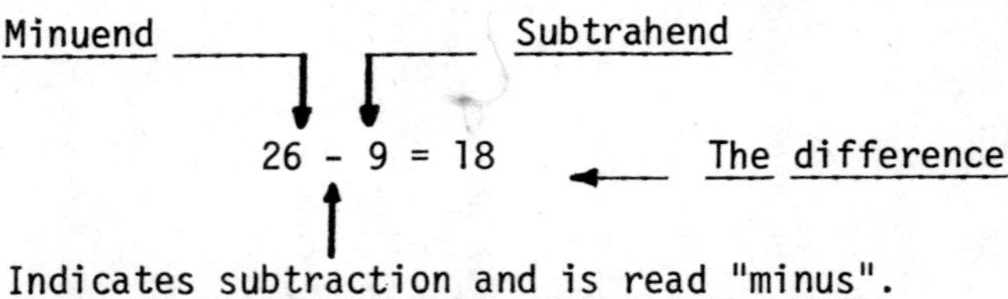

SECTION 1-3 Multiplication and Division of Whole Numbers

1. When two numbers are multiplied, the result is called the product. The numbers multiplied are called factors. The number being multiplied is called the multiplicand, and the number that indicates the number of multiplications is called the multiplier.

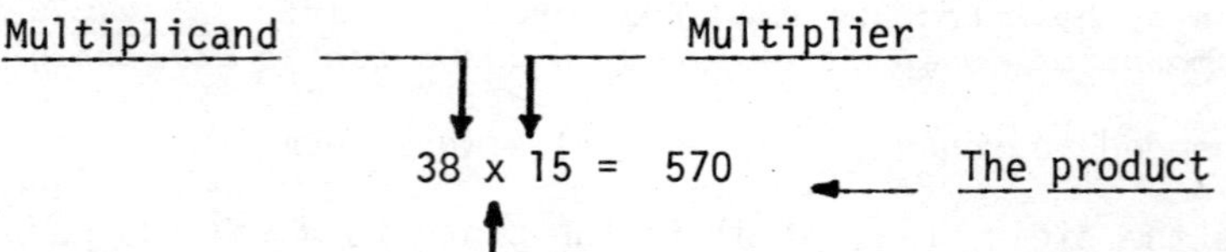

SECTION 1-3 Multiplication and Division of Whole Numbers

38 times 15 can also be written 38 · 15 and (38)(15).

2. Three properties of multiplication:

Property 1. $a \cdot 0 = 0$ and $0 \cdot a = 0$ for any number a.

The product of any number and 0 is 0.

Property 2. $a \cdot 1 = a$ and $1 \cdot a = a$ for any number a.

The product of any number and 1 is that number.

Property 3. $a \cdot b = b \cdot a$ for any numbers a and b.

The multiplicand and multiplier can be reversed and not change the product.

3. When two numbers are divided, the result is called the quotient. The number divided is called the dividend, and the number that indicates the number of divisions is called the divisor.

Dividend → 1426 ÷ 23 = 62 ← The quotient; Divisor → 23

$$1426 \div 23 = 62$$

Indicates division and is read "divided by".

1426 divided by 23 can also be written $\frac{1426}{23}$ and $23\overline{)1426}$.

4. A division problem can be checked using multiplication and addition.

quotient · divisor + remainder = dividend

5. Rule for Order of Operations

To simplify an expression with two or more operations, first do the multiplications and divisions in order from left to right, then do the additions and subtractions in order from left to right.

SECTION 1-4 Fundamental Principles of Fractions

1. A common fraction is a number that can be written as a whole number over a whole number, provided the bottom number is not zero. That is, the general form is:

$$\frac{\text{Whole number}}{\text{Whole number (not 0)}}$$

Whole number ← *The numerator*
Whole number (not 0) ← *The denominator*

2. A fraction is called proper if the numerator is less than the denominator.

$\frac{1}{2}, \frac{3}{5}, \frac{7}{10}$ *are examples of proper fractions.*

3. A fraction is called improper if the numerator is greater than the denominator.

$\frac{4}{3}, \frac{9}{7}, \frac{21}{20}$ *are examples of improper fractions.*

SECTION 1-4 Fundamental Principles of Fractions

4. A fraction is equal to 1 if the numerator and denominator are the same number, other than zero.

   $\frac{8}{8}, \frac{15}{15}, \frac{36}{36}$ *are examples of fractions equal to 1.*

5. A common fraction is reduced when numerator and denominator have no common factors other than 1.

   For example, $\frac{20}{28}$ *is not reduced, since* $\frac{20}{28} = \frac{4 \cdot 5}{4 \cdot 7}$*, and 4 is a common factor. Since 4/4 = 1, and a • 1 = a, the reduced form of the fraction is 5/7.*

SECTION 1-5 Addition and Subtraction of Fractions

1. Fractions that have the same denominator are called like fractions.

   Rule 1 states that like fractions can be added or subtracted by adding or subtracting (as indicated) the numerators of the fractions.

2. Rule 1. Adding and subtracting like fractions

   If a, b and c are whole numbers and c is not zero, then

   $\frac{a}{c} + \frac{b}{c} = \frac{a + b}{c}$ *Add the numerator and write the sum over the common denominator.*

   $\frac{a}{c} - \frac{b}{c} = \frac{a - b}{c}$ *Subtract the numerators and write the difference over the common denominator.*

3. Fractions that have different denominators are called unlike fractions.

   Unlike fractions cannot be added or subtracted until they are changed to fractions with the same denominators without changing their values. The same denominator is called the common denominator, and the smallest common denominator is called the LCD (Least Common Denominator).

4. The following three steps can be used to add or subtract unlike fractions:

   Step 1. Find the LCD of the fractions.

   Step 2. Change each fraction to one with the same value, but the LCD as denominator.

   Step 3. Find the indicated sum or difference of the like fractions obtained in Step 2.

   Step 2 is accomplished by multiplying any fraction by one, and one has the form of a whole number over a whole number, such as $\frac{2}{2}$*, or* $\frac{5}{5}$*, or* $\frac{12}{12}$.

5. A mixed number is an indicated sum of a whole number and a proper fraction.

   $2\frac{3}{4}$ *means* $2 + \frac{3}{4}$

   Whole number ↑ ↑ *Proper fraction*

SECTION 1-6 Multiplication and Division of Fractions

1. Rule 2. Multiplying common fractions

   If a, b, c and d are whole numbers (b and d are not zero), then

   $$\frac{a}{b} \cdot \frac{c}{c} = \frac{a \cdot c}{b \cdot d} \qquad \frac{\textit{Multiply the numerators}}{\textit{Multiply the denominators}}$$

   Notice that the product of common fractions does not require that the fractions have common denominators. Simply write the product of the numerators over the product of the denominators, and reduce if possible.

2. Rule 2 can be used to multiply mixed numbers by first changing the mixed numbers to improper fractions.

3. Rule 3. Dividing common fractions

   If a, b, c and d are whole numbers (b, c and d are not zero), then

   $$\frac{a}{b} \div \frac{c}{d} = \frac{a}{b} \cdot \frac{d}{c} = \frac{a \cdot d}{b \cdot c}$$

   Rule 3 is frequently stated as follows:

   "When dividing by the common fraction c/d, invert the divisor and change the division to multiplication".

PART B Selected Solutions

Exercise Set 1-1 Reading and Writing Whole Numbers

1. In the number 508,692, the 8 is in the thousands position.

5. In the number 508,692, the 0 is in the ten-thousands position.

9. In the number 137,650,924, the 1 is in the hundred-millions position.

13. In the number 35,406,982, the 3 is in the ten millions position.

17. In the number 7,633,423,098, the 8 is in the ones position.

21. 357 is written three hundred fifty-seven.

25. 360,700 is written three hundred sixty thousand, seven hundred.

29. 47,035,000 is written forty-seven million, thirty-five thousand.

33. Twelve thousand, nine hundred is written 12,900.

37. Nine million, eight hundred three thousand is written 9,803,000.

41. To the nearest hundred, 3,425 = 3,400.

    The 4 is in the hundreds position. A 2 is on the right, thus the number is rounded down.

45. To the nearest ten-thousand, 305,613 = 310,000.

    The 0 is in the ten-thousands position, a 5 is on the right, thus the number is rounded up.

Exercise Set 1-1 Reading and Writing Whole Numbers

49. To the nearest hundred-thousand, 7,160,800 = 7,200,000.

The 1 is in the hundred-thousands position. A 6 is on the right, thus the number is rounded up.

53. To the nearest million, 44,706,020 = 45,000,000.

A 4 is in the millions position. A 7 is on the right, thus the number is rounded up.

Exercise Set 1-2 Addition and Subtraction of Whole Numbers

1.
$$\begin{array}{r} {\scriptstyle 1} \\ 36 \\ 52 \\ 21 \\ +\ 48 \\ \hline 157 \end{array}$$

In the ones position, 6 + 2 + 1 + 8 = 17 and the 1 is added to the digits in the tens position.

5.
$$\begin{array}{r} {\scriptstyle 1\ \ 11} \\ 6{,}053 \\ 2{,}813 \\ 5{,}609 \\ +\ 1{,}190 \\ \hline 15{,}665 \end{array}$$

In the ones position, 3 + 3 + 9 + 0 = 15
In the tens position, 1 + 5 + 1 + 0 + 9 = 16
In the hundreds position, 1 + 0 + 8 + 6 + 1 = 16
In the thousands position, 1 + 6 + 2 + 5 + 1 = 15

9.
$$\begin{array}{r} {\scriptstyle 12\ \ 2} \\ 15{,}082 \\ 7{,}651 \\ 980 \\ +\ 40{,}815 \\ \hline 64{,}528 \end{array}$$

In the ones position, 2 + 1 + 0 + 5 = 8
In the tens position, 8 + 5 + 8 + 1 = 22
In the hundreds position, 2 + 0 + 6 + 9 + 8 = 25
In the thousands position, 2 + 5 + 7 + 0 = 14
In the ten-thousands position, 1 + 1 + 4 = 6

13. 5,945 + 684 + 13,620 + 1,925 = 22,174

Ones position: 5 + 4 + 0 + 5 = 14
Tens position: 4 + 8 + 2 + 2 + 1 = 17
Hundreds position: 9 + 6 + 6 + 9 + 1 = 31
Thousands position: 5 + 3 + 1 + 3 = 12
Ten-thousands position: 1 + 1 = 2

17. 901 + 773 + 570 + 415 + 808 + 451 = 3,918

Ones position: 1 + 3 + 0 + 5 + 8 + 1 = 18
Tens position: 0 + 7 + 7 + 1 + 0 + 5 + 1 = 21
Hundreds position: 9 + 7 + 5 + 4 + 8 + 4 + 2 = 39

21.
$$\begin{array}{r} 857 \\ -\ 405 \\ \hline 452 \end{array}$$

Ones position: 7 - 5 = 2
Tens position: 5 - 0 = 5
Hundreds position: 8 - 4 = 4

25.
$$\begin{array}{r} {\scriptstyle 7\ {}^{1}1\ {}^{1}} \\ \not{8}\ \not{2}\ 5 \\ -\ 4\ 6\ 7 \\ \hline 3\ 5\ 8 \end{array}$$

Ones position: 15 - 7 = 8
Tens position: 11 - 6 = 5
Hundreds position: 7 - 4 = 3

Exercise Set 1-2 Addition and Subtraction of Whole Numbers

29.
$$\begin{array}{r} {}^{4}\ {}^{10}\ {}^{1} \\ 13,\cancel{5}\cancel{1}0 \\ -\ 7,245 \\ \hline 6,265 \end{array}$$

Ones position: 10 − 5 = 5
Tens position: 10 − 4 = 6
Hundreds position: 4 − 2 = 2
Thousands position: 13 − 7 = 6

33.
$$\begin{array}{r} {}^{2}\ {}^{1}{}^{9}\qquad {}^{5}\ {}^{1}{}^{9}\ {}^{1} \\ \cancel{3},\cancel{0}48,\cancel{6}\cancel{0}0 \\ -\ 572,315 \\ \hline 2,476,285 \end{array}$$

Ones position: 10 − 5 = 5
Tens position: 10(changed to 9) − 1 = 8
Hundreds position: 5 − 3 = 2
Thousands position: 8 − 2 = 6
Ten-thousands position: 14 − 7 = 7
Hundred-thousands position: 10(changed to 9) − 5 = 4
Millions position: 2 − 0 = 2

37.
$$\begin{array}{r} 426 \\ +\ 398 \\ \hline 824 \end{array} \quad \text{and} \quad \begin{array}{r} 1,082 \\ -\ 824 \\ \hline 258 \end{array}$$

The third number is 258.

41. Difference + Subtrahend = Minuend

136 + 482 = 618

45. In the hits (H) column:

36 + 30 + 34 + 35 + 30 + 45 + 40 + 23 + 37 + 31 + 34 + 23 = 398

Exercise Set 1-3 Multiplication and Division of Whole Numbers

1.
$$\begin{array}{r} 84 \\ \times\ 7 \\ \hline 588 \end{array}$$

4 x 7 = 28
8 x 7 + 2 = 56 + 2 = 58

5.
$$\begin{array}{r} 706 \\ \times\ 9 \\ \hline 6354 \end{array}$$

6 x 9 = 54
0 x 9 + 5 = 0 + 5 = 5
7 x 9 = 63

9.
$$\begin{array}{r} 73,415 \\ \times\ 6 \\ \hline 440,490 \end{array}$$

5 x 6 = 30
1 x 6 + 3 = 6 + 3 = 9
4 x 6 = 24
3 x 6 + 2 = 18 + 2 = 20
7 x 6 + 2 = 42 + 2 = 44

13.
$$\begin{array}{r} 947 \\ \times\ 59 \\ \hline 8523 \\ 4735 \\ \hline 55,873 \end{array}$$

17.
$$\begin{array}{r} 241 \\ \times\ 382 \\ \hline 482 \\ 1928 \\ 723 \\ \hline 92,062 \end{array}$$

Exercise Set 1-3 Multiplication and Division of Whole Numbers

21.
```
   56
 x 90
5,040
```
Multiply: 56 x 9 = 504
Since 90 has one zero, affix one zero to 504 and get 5040.

25.
```
   620
 x 300
186,000
```
Multiply: 62 x 3 = 186
Since three zeros were deleted, affix three zeros to 186 and get 186,000.

29.
```
    17,300
  x  4,200
72,660,000
```
Multiply: 173 x 42 = 7266
Since four zeros were deleted, affix four zeros to 7266 and get 72,660,000

33. 235 + 25 = 260 (the multiplicand)

260 x 235 = 61,100

37.
```
   76 R2
5 ) 382
    35
     32
     30
      2
```
38 ÷ 5 = 7

32 ÷ 5 = 6

2 is less than 5

41.
```
     58
37 ) 2146
     185
      296
      296
```
214 ÷ 37 = 5

296 ÷ 37 = 8

45.
```
     146
36 ) 5256
     36
     165
     144
      216
      216
```
52 ÷ 36 = 1

165 ÷ 36 = 4

216 ÷ 36 = 6

49.
```
        302
56 ) 16,912
     16 8
        112
        112
```
169 ÷ 56 = 3

11 ÷ 56 = 0 and 112 ÷ 56 = 2

53.
```
          63
294 ) 18,522
      17 64
         882
         882
```
1852 ÷ 294 = 6

882 ÷ 294 = 3

57.
```
         418
183 ) 76,494
      73 2
       3 29
       1 83
       1 464
       1 464
```
764 ÷ 183 = 4

329 ÷ 183 = 1

1464 ÷ 183 = 8

Exercise Set 1-3 Multiplication and Division of Whole Numbers

61. Dividend = Quotient x Divisor

Dividend = 263 x 42

= 11,046

65.

$$\begin{array}{r} 150 \text{ feet} \\ \times\ 15 \text{ cents/foot} \\ \hline 2250 \text{ cents} \end{array} \qquad \begin{array}{r} 96 \text{ feet} \\ \times\ 26 \text{ cents/foot} \\ \hline 2496 \text{ cents} \end{array}$$

Total cost = 2250 cents + 2496 cents

= $47.46

69. 18 - 4 x 2 + 30 ÷ 6

= 18 - 8 + 5 *First do the multiplications and divisions.*

= 10 + 5 *Add and subtract from left to right.*

= 15

73. 10 + 15 - 3 x 5 - 72 ÷ 9

= 10 + 15 - 15 - 8 *First do the multiplications and divisions.*

= 25 - 15 - 8 *Add and subtract from left to right.*

= 10 - 8

= 2

77. 15 - 60 ÷ 4 + 24 - 6 x 4

= 15 - 15 + 24 - 24 *First do the multiplications and divisions.*

= 0 + 24 - 24 *Add and subtract from left to right.*

= 24 - 24

= 0

Exercise Set 1-4 Fundamental Principles of Fractions

1. $\frac{3}{4}$
   a. The denominator divides a whole into 4 equal parts.
   b. The numerator specifies 3 of the parts.

5. $\frac{7}{2}$
   a. The denominator divides a whole into 2 equal parts.
   b. The numerator specifies 7 of these parts.

9. three-fifths is written $\frac{3}{5}$, or 3/5.

13. seven-fourths is written $\frac{7}{4}$, or 7/4.

Exercise Set 1-4 Fundamental Principles of Fractions

17. Since 2 is less than 3, $\frac{2}{3}$ is a proper fraction.

21. Since 20 is less than 21, $\frac{20}{21}$ is a proper fraction.

25. $\frac{3}{6} = \frac{1 \cdot \cancel{3}}{2 \cdot \cancel{3}} = \frac{1}{2}$

29. $\frac{15}{35} = \frac{3 \cdot \cancel{5}}{7 \cdot \cancel{5}} = \frac{3}{7}$

33. $\frac{24}{16} = \frac{3 \cdot \cancel{8}}{2 \cdot \cancel{8}} = \frac{3}{2}$

37. $\frac{54}{144} = \frac{3 \cdot \cancel{18}}{8 \cdot \cancel{18}} = \frac{3}{8}$

Notice, the fractions in 33 and 37 can be reduced in steps by removing smaller common factors.

33. $\frac{24}{26} = \frac{12 \cdot \cancel{2}}{8 \cdot \cancel{2}} = \frac{6 \cdot \cancel{2}}{4 \cdot \cancel{2}} = \frac{3 \cdot \cancel{2}}{2 \cdot \cancel{2}} = \frac{3}{2}$

37. $\frac{54}{144} = \frac{27 \cdot \cancel{2}}{72 \cdot \cancel{2}} = \frac{9 \cdot \cancel{3}}{24 \cdot \cancel{3}} = \frac{3 \cdot \cancel{3}}{8 \cdot \cancel{3}} = \frac{3}{8}$

Exercise Set 1-5 Addition and Subtraction of Fractions

1. $\frac{4}{3} + \frac{2}{3}$ *Like fractions*

$= \frac{4 + 2}{3}$ *Add numerators*

$= \frac{6}{3}$ *Simplify*

$= 2$ *Reduce*

5. $\frac{7}{20} + \frac{5}{20} + \frac{13}{20}$

$= \frac{7 + 5 + 13}{20}$

$= \frac{25}{20}$

$= \frac{5}{4}$

9. $\frac{5}{32} + \frac{17}{32} + \frac{4}{32}$

$= \frac{5 + 17 + 4}{32}$

$= \frac{26}{32}$

$= \frac{13}{16}$

13. The LCD is 24.

$\frac{5}{6} = \frac{5 \cdot 4}{6 \cdot 4} = \frac{20}{24}$

$\frac{3}{8} = \frac{3 \cdot 3}{8 \cdot 3} = \frac{9}{24}$

17. The LCD is 36.

$\frac{23}{18} = \frac{23 \cdot 2}{18 \cdot 2} = \frac{46}{36}$

$\frac{13}{12} = \frac{13 \cdot 3}{12 \cdot 3} = \frac{39}{36}$

21. The LCD is 14.

$\frac{1}{2} = \frac{1 \cdot 7}{2 \cdot 7} = \frac{7}{14}$ *The smallest number that 2 and 7 will divide evenly is 14.*

$\frac{6}{7} = \frac{6 \cdot 2}{7 \cdot 2} = \frac{12}{14}$

25. The LCD is 42.

$\frac{3}{2} = \frac{3 \cdot 21}{2 \cdot 21} = \frac{63}{42}$

29. The LCD is 24.

$\frac{2}{3} = \frac{2 \cdot 8}{3 \cdot 8} = \frac{16}{24}$

$\frac{1}{7} = \frac{1 \cdot 6}{7 \cdot 6} = \frac{6}{42}$

$\frac{1}{3} = \frac{1 \cdot 14}{3 \cdot 14} = \frac{14}{42}$

$\frac{3}{8} = \frac{3 \cdot 3}{8 \cdot 3} = \frac{9}{24}$

$\frac{5}{6} = \frac{5 \cdot 4}{6 \cdot 4} = \frac{20}{24}$

33. $\frac{3}{4} - \frac{5}{9}$

$= \frac{27}{36} - \frac{20}{36}$ *The LCD is 36.*

$= \frac{27 - 20}{36}$ *Subtract numerators.*

$= \frac{7}{36}$ *Simplify.*

37. $\frac{11}{15} - \frac{6}{25}$

$= \frac{55}{75} - \frac{18}{75}$ *The LCD is 75.*

$= \frac{55 - 18}{75}$ *Subtract numerators.*

$= \frac{37}{75}$ *Simplify.*

41. $\frac{5}{24} - \frac{7}{36}$

$= \frac{15}{72} - \frac{14}{72}$ *The LCD is 72.*

$= \frac{15 - 14}{72}$ *Subtract numerators.*

$= \frac{1}{72}$ *Simplify.*

45. $\frac{3}{4} - \frac{5}{9}$

$= \frac{27}{36} - \frac{20}{36}$ *The LCD is 36.*

$= \frac{27 - 20}{36}$ *Subtract numerators.*

$= \frac{7}{36}$ *Simplify.*

49. $\frac{11}{12} - \frac{3}{8}$

$= \frac{22}{24} - \frac{9}{24}$

$= \frac{13}{24}$

53. $4\frac{1}{6} = 4 + \frac{1}{6}$

$= \frac{24}{6} + \frac{1}{6}$

$= \frac{25}{6}$

57. $10\frac{2}{3} = 10 + \frac{2}{3}$

$= \frac{30}{3} + \frac{2}{3}$

$= \frac{32}{3}$

61. $\frac{11}{7}$ *Divide 11 by 7.*

$= 1 + \frac{4}{7}$ *Quotient is 1. Remainder is 4.*

$= 1\frac{4}{7}$ *As a mixed number.*

65. $\frac{27}{10}$ *Divide 27 by 10.*

$= 2 + \frac{7}{10}$ *Quotient is 2. Remainder is 7.*

$= 2\frac{7}{10}$ *As a mixed number.*

69. $3\frac{5}{6} - 1\frac{1}{7}$

$= \frac{23}{6} - \frac{8}{7}$

$= \frac{161}{42} - \frac{48}{42}$

$= \frac{113}{42}$ or $2\frac{29}{42}$

73. $4\frac{3}{8} - \frac{5}{12}$

$= \frac{35}{8} - \frac{5}{12}$

$= \frac{105}{24} - \frac{10}{24}$

77. $6\frac{2}{3} - 1\frac{3}{4}$

$= \frac{20}{3} - \frac{7}{4}$

$= \frac{80}{12} - \frac{21}{12}$

Exercise Set 1-5 Addition and Subtraction of Fractions

$= \frac{95}{24}$ or $3\frac{23}{24}$

$= \frac{59}{12}$ or $4\frac{11}{12}$

81. $24 - 1\frac{7}{8}$

$= \frac{192}{8} - \frac{15}{8}$

$= \frac{177}{8}$ or $22\frac{1}{8}$

At the start of the trading day the price of one stock was $\$22\frac{1}{8}$.

Exercise Set 1-6 Multiplication and Division of Fractions

1. $\frac{3}{4} \cdot \frac{1}{2}$ $\qquad \frac{a}{b} \cdot \frac{c}{d}$

$= \frac{3 \cdot 1}{4 \cdot 2}$ $\qquad = \frac{a \cdot c}{b \cdot d}$

$= \frac{3}{8}$ $\qquad = \frac{ac}{bd}$

5. $\frac{5}{12} \cdot \frac{6}{11}$

$= \frac{5 \cdot \overset{1}{\cancel{6}}}{\underset{2}{\cancel{12}} \cdot 11}$ $\qquad \frac{6 \div 6 = 1}{12 \div 6 = 2}$

$= \frac{5}{22}$

9. $\frac{12}{13} \cdot \frac{9}{20}$

$= \frac{\overset{3}{\cancel{12}} \cdot 9}{13 \cdot \underset{5}{\cancel{20}}}$ $\qquad \frac{12 \div 4 = 3}{20 \div 4 = 5}$

$= \frac{27}{65}$ $\qquad \frac{3 \cdot 9 = 27}{13 \cdot 5 = 65}$

13. $\frac{3}{7} \cdot \frac{7}{12}$

$= \frac{\overset{1}{\cancel{3}} \cdot \overset{1}{\cancel{7}}}{\underset{1}{\cancel{7}} \cdot \underset{4}{\cancel{12}}}$ $\qquad \frac{\begin{matrix} 3 \div 3 = 1 \\ 7 \div 7 = 1 \end{matrix}}{\begin{matrix} 7 \div 7 = 1 \\ 12 \div 3 = 4 \end{matrix}}$

$= \frac{1}{4}$

17. $\frac{15}{26} \cdot \frac{13}{20}$

$= \frac{\overset{3}{\cancel{15}} \cdot \overset{1}{\cancel{13}}}{\underset{2}{\cancel{26}} \cdot \underset{4}{\cancel{20}}}$ $\qquad \frac{\begin{matrix} 15 \div 5 = 3 \\ 13 \div 13 = 1 \end{matrix}}{\begin{matrix} 26 \div 13 = 2 \\ 20 \div 5 = 4 \end{matrix}}$

$= \frac{3}{8}$

21. $\frac{81}{100} \cdot \frac{5}{27}$

$= \frac{\overset{3}{\cancel{81}} \cdot \overset{1}{\cancel{5}}}{\underset{20}{\cancel{100}} \cdot \underset{1}{\cancel{27}}}$ $\qquad \frac{\begin{matrix} 81 \div 27 = 3 \\ 5 \div 5 = 1 \end{matrix}}{\begin{matrix} 100 \div 5 = 20 \\ 27 \div 27 = 1 \end{matrix}}$

$= \frac{3}{20}$

25. $\frac{1}{6} \cdot \frac{21}{23} \cdot \frac{4}{35}$

$= \frac{1 \cdot \overset{\cancel{3}}{\cancel{21}} \cdot \overset{2}{\cancel{4}}}{\underset{\cancel{3}}{\cancel{6}} \cdot 23 \cdot \underset{5}{\cancel{35}}}$

$= \frac{2}{115}$

29. $\frac{1}{7} \cdot \frac{8}{9} \cdot \frac{3}{4} \cdot \frac{3}{10}$

$= \frac{1 \cdot \overset{2}{\cancel{8}} \cdot \overset{1}{\cancel{3}} \cdot \overset{1}{\cancel{3}}}{7 \cdot \underset{\cancel{3}}{\cancel{9}} \cdot \underset{1}{\cancel{4}} \cdot \underset{5}{\cancel{10}}}$

$= \frac{1}{35}$

33. $12\frac{3}{8} \cdot 5\frac{1}{6}$

$= \frac{99}{8} \cdot \frac{31}{6}$

$= \frac{\overset{33}{\cancel{99}} \cdot 31}{8 \cdot \underset{2}{\cancel{6}}}$

37. $10 \cdot 4\frac{1}{2}$

$= 10 \cdot \frac{9}{2}$

$= \frac{\overset{5}{\cancel{10}} \cdot 9}{\underset{1}{\cancel{2}}}$

Exercise Set 1-6 Multiplication and Division of Fractions

$= \frac{1023}{16}$ or $63\frac{15}{16}$

$= 45$

41. $\frac{5}{16} \div \frac{10}{21}$ $\qquad \frac{a}{b} \div \frac{c}{d}$

$= \frac{5}{16} \cdot \frac{21}{10}$ $\qquad = \frac{a}{b} \cdot \frac{d}{c}$

$= \frac{\overset{1}{\cancel{5}} \cdot 21}{16 \cdot \underset{2}{\cancel{10}}}$ $\qquad = \frac{a \cdot d}{b \cdot c}$

$= \frac{21}{32}$ $\qquad$ *Reduced*

45. $\frac{13}{20} \div \frac{9}{10}$

$= \frac{13}{20} \cdot \frac{10}{9}$

$= \frac{13 \cdot \overset{1}{\cancel{10}}}{\underset{2}{\cancel{20}} \cdot 9}$

$= \frac{13}{18}$

49. $\frac{7}{64} \div \frac{14}{32}$

$= \frac{7}{64} \cdot \frac{32}{14}$

$= \frac{\overset{1}{\cancel{7}} \cdot \overset{1}{\cancel{32}}}{\underset{2}{\cancel{64}} \cdot \underset{2}{\cancel{14}}}$

$= \frac{1}{4}$

53. $12 \div \frac{8}{9}$

$= 12 \cdot \frac{9}{8}$

$= \frac{\overset{3}{\cancel{12}} \cdot 9}{\underset{2}{\cancel{8}}}$

$= \frac{27}{2}$ or $13\frac{1}{2}$

57. $\frac{20}{21} \div 8$

$= \frac{20}{21} \cdot \frac{1}{8}$

$= \frac{\overset{5}{\cancel{20}}}{21 \cdot \underset{2}{\cancel{8}}}$

$= \frac{5}{42}$

61. $2\frac{1}{4} \div 1\frac{5}{6}$

$= \frac{9}{4} \div \frac{11}{6}$

$= \frac{9 \cdot \overset{3}{\cancel{6}}}{\underset{2}{\cancel{4}} \cdot 11}$

$= \frac{27}{22}$ or $1\frac{5}{22}$

65. $6\frac{7}{12} \div 2\frac{1}{2}$

$= \frac{79}{12} \div \frac{5}{2}$

$= \frac{79 \cdot \overset{1}{\cancel{2}}}{\underset{6}{\cancel{12}} \cdot 5}$

$= \frac{79}{30}$ or $2\frac{19}{30}$

69. $10\frac{9}{16} \div 1\frac{5}{8}$

$= \frac{169}{16} \div \frac{13}{8}$

$= \frac{\overset{13}{\cancel{169}} \cdot \overset{1}{\cancel{8}}}{\underset{2}{\cancel{16}} \cdot \underset{1}{\cancel{13}}}$

$= \frac{13}{2}$ or $6\frac{1}{2}$

73. $(29\frac{1}{4})(2) + (16\frac{1}{4})(5)$

$= \frac{117}{4} \cdot \frac{2}{1} + \frac{65}{4} \cdot \frac{5}{1}$

$= \frac{234}{4} + \frac{325}{4}$

$= \frac{559}{4} = \$139\frac{3}{4}$

77. $6936 \div \frac{289}{16}$

$= \overset{24}{\cancel{6936}} \cdot \frac{16}{\underset{1}{\cancel{289}}}$

$= 384$ tiles

PART C Chapter One Sample Test

1. Write in number form:

   five million, seventy-two thousand, six hundred twenty.

   1. ______

2. Write in word form:

   25,003,960. 2. ______

In 3 and 4, round 647,109,320 to the stated place value.

3. To the nearest ten-thousand 3. ______

4. To the nearest million 4. ______

In 5-20, do the indicated operations.

5. 4,605 + 982 + 12,400 + 3,089 5. ______

6. 14,780 + 6,250 + 962 + 7,088 6. ______

7. 5,627 − 2,819 7. ______

8. 205,700 − 82,175 8. ______

9. 163 x 85 9. ______

10. 42,000 x 370 10. ______

11. 9,798 ÷ 23 11. ______

12. $\frac{1}{18} + \frac{7}{18} + \frac{13}{18}$ 12. ______

13. $\frac{7}{12} + \frac{5}{6} - \frac{3}{8}$ 13. ______

14. $\frac{5}{8} \times \frac{12}{25} \times \frac{10}{7}$ 14. ______

15. $\frac{1}{3} \div \frac{5}{9}$ 15. ______

16. $\frac{3}{15} \div \frac{5}{9}$ 16. ______

17. $\frac{3}{10} \times 1\frac{1}{9}$ 17. ______

18. $1\frac{3}{4} + 2\frac{1}{6}$ 18. ______

Chapter One Sample Test

19. $2 \frac{1}{12} - \frac{5}{9}$ 19. ______

20. $\frac{3}{8} \div \frac{1}{6}$ 20. ______

21. If 12 bottles of detergent weigh 384 ounces, how much does one bottle weigh? 21. ______

22. Kim Lee bought 150 twenty-two cent stamps, and 25 thirty cent stamps. How much did she pay for the stamps? 22. ______

23. To build a support, Hank Tesher nailed together six planks, each $1 \frac{3}{4}$ inches thick. How thick was the support that Hank made? 23. ______

24. Kate Morgan gets about $25 \frac{1}{2}$ miles per gallon on her small pick-up truck. Last week she drove 612 miles. Approximately how many gallons of gasoline did she use? 24. ______

25. Fred Deeter bought 6 shares of stock for $\$7 \frac{1}{8}$ per share and 10 shares of another stock for $\$3 \frac{7}{8}$ per share. How much did Fred pay for all the stock? 25. ______

CHAPTER TWO

DECIMALS

PART A Summary of Topics

OBJECTIVES

1. *Write a decimal fraction in word form.*
2. *Write a decimal fraction in number form.*
3. *Round off a decimal fraction to an indicated place value.*
4. *Use money as an example of decimal fractions.*
5. *Add decimal fractions.*
6. *Subtract decimal fractions.*
7. *Multiply decimal fractions.*
8. *Divide decimal fractions.*
9. *Write decimal fractions as common fractions.*
10. *Write common fractions as decimal fractions.*
11. *Learn the metric system as an application of decimal fractions.*

SECTION 2-1 Reading and Rounding Decimal Fractions

1. A dot, called a decimal point, is used to write numbers as decimal fractions.

   A decimal fraction is another way of writing common fractions whose denominators are 10, 100, 1,000, 10,000, and so on. To illustrate,

   $\frac{3}{10}$ *is written 0.3 and is read "three-tenths".*

   $\frac{7}{100}$ *is written 0.07 and is read "seven-hundredths".*

   $\frac{23}{1000}$ *is written 0.023 and is read "twenty-three thousandths".*

   Notice that the number of places in the decimal form is the same as the number of zeros in the denominator of the common fraction form.

2. The names of the first eight places for decimal fractions are listed below:

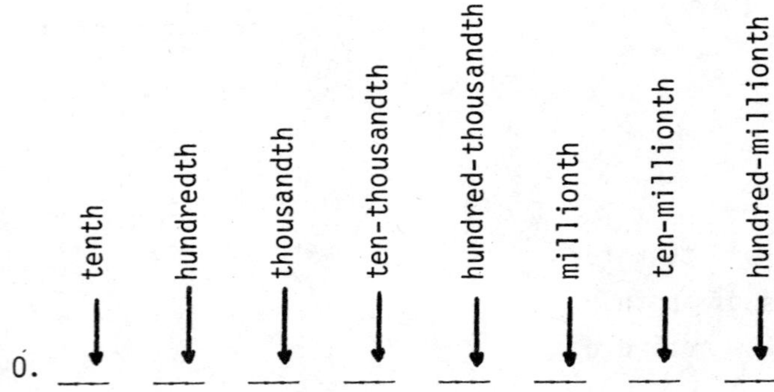

3. A decimal fraction is written in number form by using the digits and place value.

4. A decimal fraction can be written in word form by writing the number in words as it is read.

   The suffix, th , distinguishes between words for place values to the left of the decimal point and place values to the right of the decimal point.

ten-thousand - fifth place to the left of decimal point
ten-thousandth - fourth place to the right of decimal point

5. Two steps are needed to round (or round-off) a decimal fraction:

   Step 1. Locate the digit (call it d) in the place to which the number is to be rounded.

   Step 2. Look only at the digit immediately to the right of d.

   a. If this digit is 0, 1, 2, 3 or 4, then change all the digits to the right of d to zeros.

   b. If this digit is 5, 6, 7, 8 or 9, then add one to d and change all the digits to the right of d to zeros.

6. In the United States monetary system, the whole number part is read dollar and the first two places of the decimal part are read cent.

SECTION 2-2 Addition and Subtraction of Decimals

1. To add decimal fractions, the digits in the same place values are added.

   To help determine which digits are in the same place values, the numbers are put in a column with the decimal points aligned. The numbers can then be added as though they were whole numbers.

2. To subtract decimal fractions, the digits in the same place values are subtracted.

   Again, the decimal points should be aligned to subtract decimal fractions. It may be necessary to supply zeros to the minuend, if the subtrahend has more decimal places.

5.300	*Write 5.3 as 5.300 to get three decimal places.*
- 2.168	*The subtrahend has three decimal places.*

3. As in the case of whole numbers, a difference of two decimal fractions can be checked with addition.

minuend	*15.60*	*Check*	*subtrahend*	*8.28*
subtrahend	*- 8.28*		*difference*	*+ 7.32*
difference	*7.32*		*minuend*	*15.60*

4. Money amounts in the United States system are decimal fractions. Therefore adding and subtracting money amounts may require adding and subtracting decimal fractions.

SECTION 2-3 Multiplication of Decimals

1. Rule 1 can be used to locate the decimal point in the product of decimal fractions.

   Rule 1. The product of decimal fractions x and y has as many decimal places as the sum of the number of decimal places in x and y.

   Apart from locating the decimal point in the product, two decimal fractions are multiplied as though they were whole numbers.

SECTION 2-3 Multiplication of Decimals

2. Numbers such as 100, 10,000 and so on are powers of 10, in that they can be written as 10 with an exponent. The exponent indicates the number of zeros in the power of 10.

$100 = 10^2$ the exponent is 2 and is read "10 squared".

$1{,}000 = 10^3$ the exponent is 3 and is read "10 cubed".

$10{,}000 = 10^4$ the exponent is 4 and is read "10 to the fourth power".

3. Rule 2. To multiply a number x by a power of 10, move the decimal point in x to the right the number of places equal to the exponent.

$1.25 \times 10^3 = 1{,}250$ Move the decimal point 3 places.

$0.00068 \times 10^4 = 6.8$ Move the decimal point 4 places.

4. Rule 3. To multiply a number x by a power of 0.1, move the decimal point in x to the left the number of places equal to the exponent.

$163{,}000 \times 0.1^4 = 16.3$ Move the decimal point 4 places.

$5.75 \times 0.1^2 = 0.0575$ Move the decimal point 2 places.

SECTION 2-4 Division of Decimals

1. The following steps can be used to locate the decimal point in a quotient of two numbers when at least one of the numbers is a decimal fraction.

   Rule 4. To locate the decimal point in a quotient:

   Step 1. If necessary, move the decimal point in the divisor to the right to make it a whole number.

   Step 2. Move the decimal point in the dividend to the right the same number of places as were used in Step 1.

   Step 3. Put a decimal point in the quotient directly above the location of the decimal point in the dividend and divide.

2. It may be necessary to supply zeros to the dividend to move the decimal point to the right the required number of places in Step 2.

0.025. ⟌17.500. *Supply two zeros*

Move 3 places. *Move 3 places.*

3. If a quotient has a remainder, it may be possible to write the remainder as a decimal fraction by supplying more zeros to the dividend and continuing the dividing process.

4. In many business problems, a quotient that has a remainder is rounded to some specified place value. To round such a quotient it is necessary to compute the quotient to one more place value than the round-off position.

```
     16.892
28 ⟌473.000
```

To round this quotient to the nearest hundredth, the quotient is computed to three decimal places and rounded to two.

Thus, $473 \div 28 = 16.89$ to the nearest hundredth.

SECTION 2-5 Decimal Fractions and Common Fractions Compared

1. The following three-step procedure can be used to write a decimal fraction as a common fraction.

   Step 1. Change the decimal fraction to a whole number and write it as the numerator of the common fraction.

   Step 2. Write as the denominator a 1 followed by as many zeros as there are decimal places in the given decimal fraction.

   Step 3. If possible, reduce the common fraction formed.

2. To change any common fraction to a decimal, divide the numerator of the fraction by the denominator.

   a. For some common fractions the division will yield a zero remainder, provided enough zeros are supplied to the dividend.

      Common fractions with denominators 2, 4, 5, 8, 10, 16, 20, 25, 32, 40, 50, 64, and so on will always give a zero remainder, provided the process is continued long enough.

   b. Many common fractions will never yield a zero remainder when they are changed to decimal form. Frequently such decimal numbers are rounded to some specified number of places. As such, the decimal representation of the number is only an approximation.

      Common fractions with denominators 3, 6, 7, 9, 11, 12, 13, 14, 15, 17, and so on will never give a zero remainder, no matter how long the process is continued.

SECTION 2-6 The Metric System

1. In the metric system of measurement;

   a. the standard unit of length is the meter.

   b. the standard unit of weight is the kilogram.

2. The following abbreviations are used for the units meter, gram and liter.

   meter is abbreviated *m*
   gram is abbreviated *g*
   liter is abbreviated *l*

3. The following prefixes are used in the metric system:

   kilo (k) - means thousand   1 km = 1,000 m

   hecto (h) - means hundred   1 hm = 100 m

   deka (dk) - means ten   1 dkm = 10 m

   deci (d) - means tenth   1 dm = 0.1 m

   centi (c) - means hundredth   1 cm = 0.01 m

   milli (m) - means thousandth   1 mm = 0.001 m

4. The following steps can be used to change a given metric measurement to one with smaller units.

SECTION 2-6 The Metric System

Rule 1. To change a measurement from a larger unit to a smaller unit:

Step 1. Count the number of prefixes (or unit) from the given prefix in the measurement to the desired prefix (or unit).

Step 2. Move the decimal point in the given measure to the right the number of places equal to the count obtained in Step 1.

5. The following list contains conversions that are commonly made in the metric system.

LENGTH	*WEIGHT*	*VOLUME*
1. *kilometers to meters (km to m)*	1. *kilograms to grams (kg to g)*	1. *liters to milliliters (l to ml)*
2. *meters to centimeters (m to cm)*	2. *grams to milligrams (g to mg)*	2. *liters to cubic centimeters (l to cc)*
3. *centimeters to millimeters (cm to mm)*		

6. The following steps can be used to change a given metric measurement to one with larger units.

Rule 2. To change a measurement from a smaller unit to a larger unit:

Step 1. Count the number of prefixes (or unit) from the given prefix in the measurement to the desired prefix (or unit).

Step 2. Move the decimal point in the given measure to the left the number of places equal to the count obtained in Step 1.

7. The following list contains conversions that are commonly made in the metric system.

LENGTH	*WEIGHT*	*VOLUME*
1. *meters to kilometers (m to km)*	1. *grams to kilograms (g to kg)*	1. *milliliters to liters (ml to l)*
2. *centimeters to meters (cm to m)*	2. *milligrams to grams (mg to g)*	2. *cubic centimeters to liters (cc to l)*

8. The following tables contain some of the conversion factors used to change Metric units to equivalent English units, and vice versa.

	METRIC TO ENGLISH	ENGLISH TO METRIC
LENGTH:	*1 mm = 0.03937 inch*	*1 inch = 25.40 mm*
	1 cm = 0.3937 inch	*1 inch = 2.54 cm*
	1 m = 39.37 inches	*1 inch = 0.0254 m*
	1 m = 3.281 feet	*1 foot = 0.3048 m*
	1 m = 1.094 yards	*1 yard = 0.914 m*
	1 km = 0.6214 mile	*1 mile = 1.609 km*

SECTION 2-6 The Metric System

	METRIC TO ENGLISH	ENGLISH TO METRIC
VOLUME:	*1 ml = 0.0338 fluid ounce*	*1 fluid ounce = 29.58 ml*
	1 l = 33.81 fluid ounces	*1 fluid ounce = 0.02958 l*
	1 l = 2.113 liquid pints	*1 liquid pint = 0.4732 l*
	1 l = 1.057 liquid quarts	*1 liquid quart = 0.9464 l*
	1 l = 0.2642 liquid gallon	*1 liquid gallon = 3.785 l*

	METRIC TO ENGLISH	ENGLISH TO METRIC
WEIGHT:	*1 mg = 0.00003527 ounce*	*1 ounce = 28.350 mg*
	1 g = 0.03527 ounce	*1 ounce = 28.35 g*
	1 g = 0.002205 pound	*1 pound = 453.6 g*
	1 kg = 2.205 pounds	*1 pound = 0.4536 kg*

PART B Selected Solutions

Exercise Set 2-1 Reading and Rounding Decimal Fractions

1. In 0.720961 the 2 is in the hundredths position.

5. In 3.1415926 the 2 is in the millionths position.

9. In words, 0.48 is written forty-eight hundredths.

13. In words, 4.16 is written four and sixteen hundredths.

    Write and for the decimal point.

17. In words, 163.9 is written one hundred sixty-three and nine tenths.

21. Forty-two hundredths is written 0.42.

    The 0 is written to the left of the decimal point to emphasize that the 42 is a decimal fraction and not a whole number.

25. Seven and twenty-seven ten thousandths is written 7.0027.

    Two zeros are needed to locate the 7 in the ten-thousandths position.

29. One hundred eighteen thousandths is written 0.118.

33. To the nearest hundred thousandths, 8.670914 = 8.67091.

    The 1 is in the hundred thousandths position. The digit 4 is to the right, thus the number is rounded down.

37. To the nearest hundredths, 17.82495 = 17.82.

    The 2 is in the hundredths position. The digit 4 is to the right, thus the number is rounded down.

41. To the nearest tens, 26.06715 = 30.

    The 2 is in the tens position. Notice that the word is tens, and not tenths.

Exercise Set 2-1 Reading and Rounding Decimal Fractions

45. To the nearest hundredths, 628.05963 = 628.06.

The 5 is in the hundredths position. The 9 is to the right, therefore the number is rounded up.

49. a. To the nearest cent, \$67.9849 = \$67.98 (Rounded down)

b. To the nearest dollar, \$67.9849 = \$68 (Rounded up)

Exercise Set 2-2 Addition and Subtraction of Decimals

```
1.  13.5      In the tenths position, 5 + 9 = 14
     8.9      In the ones position, 3 + 8 + 1 = 12
    22.4      In the tens position, 1 + 1 = 2
```

```
     1 1
5.  3.065
    0.970   Supply a zero.
    4.035
```

```
     1 11
9.  0.0098
    0.0510   Supply 1 zero.
    4.9600   Supply 2 zeros.
    0.8140   Supply 1 zero.
    5.8348
```

```
      1 21
13.  0.098
     0.690
     7.040
     3.603
    11.431
```

```
     1111 1
17.  5800.92
      672.10
     1000.49
       78.30
     7551.81
```

```
         1 1
21. a.  2.790    b.  3.0
        0.431        0.4
        1.200        1.0
        4.421        4.4
```

```
          21
25. a.  0.0830    b.  0.08
        0.1900        0.2
        0.0048        0.005
        0.5690        0.6
        0.8468        0.885
```

```
         11 1
29. a.    0.89   b.  0.9
          5.33       5.
         14.20      10.
          6.10       6.
          0.36       0.4
         26.88      22.3
```

```
        41
33.   8.50
    - 5.27
      3.23
```

```
             1
            8 11
37.   0.0 9 2 5
    - 0.0 7 6 8
      0.0 1 5 7
```

```
       3 9 9 1
41.   5 4 0.0 0
    -   8 9.0 5
      4 5 0.9 5
```

```
         5 1
45.   0.6 4 8
    - 0.3 9 0
      0.2 5 8
```

```
       2 9 1
49.   13.0 0 8
    -  8.3 2 0
       4.6 8 8
```

```
53.   $39.00 Regular price
    -  29.25 Sale price
      $ 9.75
```

```
57. a. $927.34    b. $744.71    c. $708.42    d. $428.42
      - 182.63      -  36.29     - 280.00      -  93.57
       $744.71       $708.42      $428.42       $334.85
```

Exercise Set 2-2 Addition and Subtraction of Decimals

```
e.  $334.85    Balance after check number 103.   f.  $256.40
  -   78.45    Amount of check number 104.          - 133.48
    $256.40    New balance.                         $122.92
```

Exercise Set 2-3 Multiplication of Decimals

```
1.     830      0 decimal places        5.   0.029     3 decimal places
    x 0.46    + 2 decimal places           x  6800   + 0 decimal places
      4980                                   23200
     3320                                   174
    381.80      2 decimal places            197.200    3 decimal places
```

```
9.   160.05     2 decimal places       13.    1.08     2 decimal places
   x  20.04   + 2 decimal places            x 0.005  + 3 decimal places
      64020                                0.00540     5 decimal places
    3201000
    3207.4020   4 decimal places
```

```
17.    1.2      1 decimal place
    x 0.03    + 2 decimal places
     0.036      3 decimal places
      0.09    + 2 decimal places
   0.00324      5 decimal places
```

21. 1.96×100

$= 1.96 \times 10^2$ *Two zeros*

$= 196$ *Move two places.*

25. $0.0082 \times 10{,}000$

$= 0.0082 \times 10^4$ *Four zeros*

$= 82$ *Move four places.*

29. $0.00051 \times 1{,}000{,}000$

$= 0.00051 \times 10^6$ *Six zeros*

$= 510$ *Move six places.*

```
33.  $20.50 each set          37.   $9.95  Bath towel
     x    3 sets                  x  0.06  tax rate
     $61.50 Total cost            $0.5970
```

= $0.60, to the nearest cent

Exercise Set 2-4 Division of Decimals

```
        7.2                         0.68
1.  6 ) 43.2               5.  21 ) 14.28
        42                          12 6
         12                          1 68
         12                          1 68
```

```
          0.0035
9.  28 ) 0.0980
           84
           140
           140
```

13. $0.4648 \div 0.83 = 46.48 \div 83$

```
        0.56
   83 ) 46.48
        41 5
         4 98
         4 98
```

Exercise Set 2-4 Division of Decimals

17. $0.2016 \div 0.028$

$= 201.6 \div 28$

```
      7.2
 28 ) 201.6
      196
        5 6
        5 6
```

21.

```
     12.25
 8 ) 98.00
     8
     18
     16
      20
      16
       40
       40
```

Supply two zeros and continue dividing.

25.

```
      0.082
 25 ) 2.050
      2 00
         50
         50
```

Supply one zero and continue dividing.

29.

```
      0.00195
 56 ) 0.10920
         56
         532
         504
          280
          280
```

Supply one zero and continue dividing.

33. $16.73 \div 3.5$

$= 167.3 \div 35$

```
        4.78
 35 ) 167.30
      140
       273
       245
        280
        280
```

37. $0.0292 \div 0.008$

$= 29.2 \div 8$

```
     3.65
 8 ) 29.20
     24
      52
      48
       40
       40
```

In 41, 45 and 49, divide to three decimal places, then round to two decimal places.

41.

```
      2.083 = 2.08 to two decimal
 6 ) 12.500         places.
     12
       50
       48
        20
        18
```

45.

```
      3.928 = 3.93 to two decimal
 7 ) 27.500         places.
     21
      6 5
      6 3
        20
        14
         60
         56
```

49.

```
       0.704 = 0.70 to two decimal
 44 ) 31.000         places.
      30 8
        200
        176
```

53. $500 \div 1.593$

$= 313.873$

$= 313.87$ to the nearest hundredth.

57. $\$20.88 \div 24 = \$0.87$ each

Exercise Set 2-5 Decimal Fractions and Common Fractions Compared

1. $0.3 = \frac{3}{10}$ — *Write 0.3 as the whole number 3* / *Write 1 followed by one zero*

Exercise Set 2-5 Decimal Fractions and Common Fractions Compared

5. $0.1875 = \frac{1875}{10000}$ *Write 0.1875 as the whole number 1875 / Write 1 followed by four zeros*

$= \frac{3}{16}$ *Reduce the fraction.*

9. $9.4 = \frac{94}{10}$ *Write 9.4 as the whole number 94 / Write 1 followed by one zero*

$= \frac{47}{5}$ *Reduce the fraction.*

13.
```
     0.8
  5 ) 4.0
      4 0
```
Thus, $\frac{4}{5} = 0.8$

17.
```
      0.0125
  80 ) 1.0000
        80
        200
        160
         400
         400
```
Thus, $\frac{1}{80} = 0.0125$

21.
```
      1.55
  20 ) 31.00
       20
       11 0
       10 0
        1 00
        1 00
```
Thus, $\frac{31}{20} = 1.55$

25.
```
     0.6666
  3 ) 2.0000
      1 8
        20
        18
         20
         18
          20
          18
```
To three decimal places, $\frac{2}{3} = 0.667$

29.
```
      0.5333
  15 ) 8.0000
       7 5
         50
         45
          50
          45
           50
           45
```
To three decimal places, $\frac{8}{15} = 0.533$

33.
```
      1.7142
  7 ) 12.0000
      7
      5 0
      4 9
        10
         7
         30
         28
          20
          14
```
To three decimal places, $\frac{12}{7} = 1.714$

37. $3.25 = \frac{325}{100} = \frac{13}{4}$

41. $0.15 = \frac{15}{100} = \frac{3}{20}$

Exercise Set 2-6 The Metric System

1. 5 kilometers is written 5 km. *k for kilo and m for meter.*
5. 12 centimeters is written 12 cm. *c for centi and m for meter.*
9. 500 milligrams is written 500 mg. *m for milli and g for gram.*
13. 9 kg stands for 9 kilograms.
17. 9 kl stands for 9 kiloliters

Exercise Set 2-6 The Metric System

21. 0.73 km = 730 m *Move the decimal point 3 places right.*

25. 4,650 g = 4.65 kg *Move the decimal point 3 places left.*

29. 61.8 cc = 61.8 ml *One cc equals one ml.*

33. 3.02 km = 302,000 cm *Move the decimal point 5 places right.*

37. 1.035 kg = 1,035,000 mg *Move the decimal point 6 places right.*

41. 0.00075 km = 0.75 m *The decimal point is moved 3 places right.*

45. 0.023 kg = 23 g *The decimal point is moved 3 places right.*

49. 0.032 l = 32 ml *The decimal point is moved 3 places right.*

53. 5,000 cc = 5 l *The decimal point is moved 3 places left.*

57. 1,400 m = 1.4 km *The decimal point is moved 3 places left.*

61. 8(0.03937)

= 0.31496

= 0.31 inches, to two places

65. 40(2.54)

= 101.6 cm

69. 6(3.281)

= 19.686

= 19.69 feet, to two places

73. 1.8(1.609)

= 2.8962

= 2.90 km, to two places

77. 3.5(473.2) *0.4732 l*

= 1656.2 ml *= 473.2 ml*

PART C Chapter Two Sample Test

In 1 and 2, write each number in number form.

1. Eight and seventy-five hundredths 1. ______________

2. One hundred thirty-six ten thousandths 2. ______________

In 3 and 4, write each number in word form.

3. 12.09 3. ______________________________

4. 0.00953 4. ______________________________

In 5-14, do the indicated operations.

5.
```
    10.7
     0.803
     5.62
  + 21.075
```

6.
```
    3.408
  - 1.5329
```

7.
```
    13.5
  x  6.4
```

5. ______________

6. ______________

7. ______________

8. 2.035 + 15.96 + 0.927 8. ______________

9. 25.06 - 17.825 9. ______________

10. 30.4 x 0.48 10. ______________

11. 398,000 x 0.00001 11. ______________

12. 0.068 ÷ 0.4 12. ______________

13. 15 ÷ 0.003 13. ______________

14. 4.63 ÷ 0.015 (to two decimal places) 14. ______________

Chapter Two Sample Test

In 15 and 16, write each decimal fraction as a reduced common fraction.

15. 3.6 15. ________

16. 7.25 16. ________

In 17 and 18, write each common fraction as an exact decimal fraction.

17. $\frac{37}{20}$ 17. ________

18. $\frac{13}{16}$ 18. ________

In 19 and 20, write each common fraction as a decimal, rounded to three decimal places.

19. $\frac{26}{11}$ 19. ________

20. $\frac{3}{14}$ 20. ________

In 21 and 22, round 2.718936 to the indicated place value.

21. To the nearest hundredth 21. ________

22. To the nearest ten-thousandth 22. ________

23. The Niles Corporation manufactures cookware at its plant in Warren, Oregon. Plant records indicate that 7/12 of the cost of manufacturing the cookware is direct labor costs. Write 7/12 as a decimal rounded to three places. 23. ________

24. The Gunderson family maintains a budget. 0.21 of the budget is food, 0.38 of the budget is housing and housing costs, and 0.17 of the budget is related to the three cars in the family. What part of the budget is for food, housing costs and car costs? 24. ________

CHAPTER THREE

EQUATIONS IN ONE VARIABLE, RATIO AND PROPORTIONS

PART A Summary of Topics

OBJECTIVES

1. *Learn what is a solution of an equation.*
2. *Learn what is meant by solving an equation.*
3. *Learn what is the general form of a linear equation in one variable.*
4. *Learn what is meant by equivalent equations.*
5. *Learn four properties of equality that can be used to solve an equation.*
6. *Learn what is a formula.*
7. *Learn how to evaluate a formula.*
8. *Learn the definition of a ratio.*
9. *Learn the general form of a proportion.*
10. *Use the means-extremes product property of a proportion.*
11. *Solve business problems using a linear equation in one variable.*
12. *Solve business problems using a proportion.*

SECTION 3-1 Solving Equations in One Variable

1. An equation is a mathematical statement in which the verb is equals (written =) and which asserts that the expressions to the left and right of the equals sign name the same number.

   $7 + 8 = 15$ is a true equation

   $9 \cdot 3 = 27$ is a true equation

   $6 + 3 = 8$ is a false equation

   $8 \cdot 5 = 39$ is a false equation

   $x + 9 = 14$ is neither true nor false (sometimes called a conditional equation).

   When x is replaced with 5, then $5 + 9 = 14$ is true. When x is replaced with any other number, then the equation is false. To solve an equation means to find the replacements for x that result in a true statement.

2. A number replacement for a variable that makes an equation a true statement is called a solution, or root, of the equation.

   In the equation $x + 9 = 14$, the solution is 5, since $5 + 9 = 14$ is a true statement.

3. A linear equation in one variable is one that can be written in the form

$$ax + b = c$$

   where a, b, and c are numbers (a does not equal 0) and x is a variable.

   In the equation $ax + b = c$, the b and c terms are called constants, and ax is called a variable term.

4. The following properties of equality can be used to solve equations studied in this text.

SECTION 3-1 Solving Equations in One Variable

In each equation, a and b are any numbers and c is any number except 0 (written $c \neq 0$).

Addition Property of Equality

If $a = b$, then $a + c = b + c$ "Equals added to equals are equal."

Subtraction Property of Equality

If $a = b$, then $a - c = b - c$ "Equals subtracted from equals are equal."

Multiplication Property of Equality

If $a = b$, then $a \cdot c = b \cdot c$ "Equals multiplied by equals are equal."

Division Property of Equality

If $a = b$, then $\frac{a}{c} = \frac{b}{c}$ "Equals divided by equals are equal."

5. To check an equation, replace the variable with the number obtained as a solution. Then simplify both sides of the equation. If the resulting number equation is true, then the solution is accepted. If the number equation is false, then the equation must be solved again.

6. Equations (1) and (2) can be used to combine (that is, add or subtract) variable terms. In these equations a and b are numbers and x is a variable.

$$ax + bx = (a + b)x \qquad (1)$$

$$ax - bx = (a - b)x \qquad (2)$$

$$10x + 3x = (10 + 3)x = 13x$$

$$5y + y = (5 + 1)y = 6y$$ *NOTE: y means $1 \cdot y$*

$$13z - 9z = (13 - 9)z = 4z$$

$$7t - t = (7 - 1)t = 6t$$ *NOTE: t means $1 \cdot t$*

SECTION 3-2 Formulas

1. A formula is a mathematical equation in which the letters (or variables) represent number quantities from the world that are related to each other in a way described by the equation.

   As will be seen, there are many formulas that relate quantities in the business world.

2. When values are given for all but one variable in a formula, then the formula can be evaluated for the remaining variable. The following three-step procedure can be used to evaluate a formula.

   Step 1. Replace each variable in the equation with its given value.

   Step 2. Observe the rule for order of operations and simplify the resulting expressions in the equation.

   Step 3. If necessary, solve the equation in Step 2 for the remaining variable.

SECTION 3-2 Formulas

3. Many business related formulas contain grouping symbols, such as parentheses ( ), braces { }, and the fraction bar ——. The rule for order of operations must be expanded to cover expressions with grouping symbols.

    1. Simplify any expression separated by a grouping symbol.

    2. The sequence in which operations are performed is as follows:

    a. Find the value of any number with an exponent.

    b. Do the multiplications and divisions in the order in which they occur, from left to right.

    c. Do the additions and subtractions in the order in which they occur, from left to right.

In the expression 4(9 + 3), the addition is done before the multiplication, because the 9 + 3 is inside parentheses. Thus, 4(9 + 3) = 4(12) = 48. Without parentheses the multiplication is done first. That is, 4 • 9 + 3 = 36 + 3 = 39. It is very important to observe the order of operations rule when evaluating a formula.

SECTION 3-3 Ratio and Proportion

1. A ratio is a comparison of two numbers, or two similar quantities, by division.

When two similar quantities are compared, the quantities should be written in the same units. Thus, when two times are compared, the quantities should both be expressed in minutes, or hours, or days, or some other unit of time.

2. In many applied problems, two quantities with different units are compared as ratios. As such, these ratios are written with the units in numerator and denominator.

Two commonly used ratios of dissimilar units are used for cars and trucks: $\frac{\text{miles}}{\text{hour}}$ *for speed, written mph, and* $\frac{\text{miles}}{\text{gallons}}$ *for fuel economy, written mpg.*

3. A proportion is an equation that states that two ratios are equal.

If $\frac{a}{b}$ and $\frac{c}{d}$ are ratios ($b \neq 0$ and $d \neq 0$), then

$$\frac{a}{b} = \frac{c}{d}$$

is a proportion. Read the equation as follows "a is to b as c is to d." In the given equation, a, b, c, and d are called the terms of the proportion.

If the ratios $\frac{a}{b}$ *and* $\frac{c}{d}$ *compare different units, then the similar units are written in the numerators and denominators. That is, a and c have the same unit, while b and d have the same unit. To illustrate, if three cans of olives cost \$3.27, then six cans cost \$6.54. As a proportion:*

$$\frac{3\text{ cans}}{\$3.27} = \frac{6\text{ cans}}{\$6.54}$$

cans on top
dollar amount on bottom

SECTION 3-3 Ratio and Proportion

4. The means-extremes product property of a proportion.

   If $\frac{a}{b} = \frac{c}{d}$, then $a \cdot d = b \cdot c$.

   $a \cdot d$ is called the product of the extremes.

   $b \cdot c$ is called the product of the means.

   This property of proportions is sometimes called "cross-multiplication". That is, the product of the upper left term and lower right term is equal to the product of the lower left term and upper right term. The products multiply "cross-terms".

SECTION 3-4 Applications

1. Many business problems can be solved by using a linear equation in one variable. The following five-step procedure can be followed for solving such problems with a linear equation.

   Step 1. Assign a letter to the unknown quantity in the problem.

   Step 2. Write an equation using the relationship stated in the problem.

   Step 3. Solve the equation for the variable.

   Step 4. Check the solution in the context of the word problem.

   Step 5. Answer the question in the problem.

   The key step in the procedure is Step 2. In most problems, all the given information must be used to write the equation. The primary weapon for developing a knack for solving applied problems is perseverence. Keep trying, and eventually the effort will pay off.

2. Many business problems can be solved with a proportion.

   The units in the ratios of a proportion are valuable in writing a correct equation. That is, make certain the numerators have the same units, and the denominators have the same units.

PART B Selected Solutions

Exercise Set 3-1 Solving Equations in One Variable

1. $3x + 2 = 14$ and $x = 4$

$3(4) + 2 = 14$	*Replace x by 4.*
$12 + 2 = 14$	*Multiply before adding.*
$14 = 14$, true	*Thus, 4 is the solution.*

Exercise Set 3-1 Solving Equations in One Variable

5. $12 + 10z = 14z$ and $z = 3$

$12 + 10(3) = 14(3)$ *Replace z by 3.*

$12 + 30 = 42$ *Multiply before adding.*

$42 = 42$, true *Thus, 3 is the solution.*

9. $3b + 8 = 5b + 1$ and $b = 5$

$3(5) + 8 = 5(5) + 1$ *Replace b by 5.*

$15 + 8 = 25 + 1$ *Multiply before adding.*

$23 = 26$, false *Thus, 5 is not the solution.*

13. $y = 21$ and $9y = 190$

$9(21) = 190$ *Replace y by 21 in 9y = 190.*

$189 = 190$, false *Thus, 21 is not the solution.*

17. $t = 13$ and $2t + 25 = 70 - 3t$

$2(13) + 25 = 70 - 3(13)$ *Replace t by 13.*

$26 + 25 = 70 - 39$ *Multiply before adding.*

$51 = 31$, false *Thus, 13 is not the solution.*

21. $x - 5 = 8$

$x = 13$ *Add 5 to both sides.*

The solution is 13.

25. $8z = 72$

$z = 9$ *Divide both sides by 8.*

The solution is 9.

29. $5b - 12 = 38$

$5b = 50$ *Add 12 to both sides.*

$b = 10$ *Divide both sides by 5.*

The solution is 10.

33. $39 = 7x + 39$

$0 = 7x$ *Subtract 39.*

$0 = x$ *Divide by 7.*

The solution is 0.

37. $9 = 5 + \frac{2}{3}z$

$4 = \frac{2}{3}z$ *Subtract 5.*

$6 = z$ *Multiply by* $\frac{3}{2}$.

The solution is 6.

41. $2b = b + 5$

$2b - b = 5$ *Subtract b from both sides.*

$b = 5$ $2b - b = (2 - 1)b = 1 \cdot b = b$

The solution is 5.

Exercise Set 3-1 Solving Equations in One Variable

45. $3y - 2 = y + 2$

$3y = y + 4$	*Add 2 to both sides.*
$3y - y = 4$	*Subtract y from both sides.*
$2y = 4$	*$3y - y = (3 - 1)y = 2y$.*
$y = 2$	*Divide both sides by 2.*

The solution is 2.

49. $19 - 2a = 6a - 13$

$32 - 2a = 6a$	*Add 13 to both sides.*
$32 = 6a + 2a$	*Add 2a to both sides.*
$32 = 8a$	*$6a + 2a = (6 + 2)a = 8a$.*
$4 = a$	*Divide both sides by 8.*

The solution is 4.

53. Replace y by 4 in $y - \underline{?} = 3y - 12$

$4 - \underline{?} = 3(4) - 12$

$4 - \underline{?} = 12 - 12$

$4 - \underline{?} = 0$

Since $4 - 4 = 0$, the ? must be 4.

57. Replace a by 10 in $8a + 2 = \underline{?} \cdot a + 32$

$8(10) + 2 = \underline{?} \cdot 10 + 32$

$80 + 2 = \underline{?} \cdot 10 + 32$

$82 = \underline{?} \cdot 10 + 32$

$50 = \underline{?} \cdot 10$

Since $5 \cdot 10 = 50$, the ? must be 5.

Exercise Set 3-2 Formulas

1.

$P = 4s$	*The given formula.*
$P = 4(15)$	*Replace s by 15.*
$P = 60$	*Simplify the indicated product.*

5.

$A = l \cdot w$	*The given formula.*
$92 = l \cdot 4$	*Replace A by 92 and w by 4.*
$23 = l$	*Divide both sides by 4.*

Exercise Set 3-2 Formulas

9. $V = l \cdot w \cdot h$ — *The given formula.*

$1440 = 12(8)h$ — *Replace V by 1,440, l by 12, and 2 by 8.*

$1440 = 96h$ — *Multiply first.*

$15 = h$ — *Divide both sides by 96.*

13. $C = \frac{5}{9}(F - 32)$ — *The given formula.*

$C = \frac{5}{9}(212 - 32)$ — *Replace F by 212.*

$C = \frac{5}{9}(180)$ — *Subtract first.*

$C = 100$ — *$\frac{5}{9}(180) = 5(20) = 100$.*

17. $A = P + PRT$ — *The given formula.*

$1720 = 1000 + 1000(0.12)T$ — *Replace A by 1720, P by 1000, and R by 0.12.*

$1720 = 1000 + 120T$ — *Multiply first.*

$720 = 120T$ — *Subtract 1000 from both sides.*

$6 = T$ — *Divide both sides by 120.*

21. $MC\% = \frac{M}{C}$ — *The given formula.*

$MC\% = \frac{30}{150}$ — *Replace M by 30 and C by 150.*

$MC\% = \frac{1}{5}$ or 0.20 — *Reduce fraction, or change to decimal.*

25. $MD\% = \frac{MD}{RP}$ — *The given formula.*

$MD\% = \frac{9.70}{48.50}$ — *Replace MD by 9.70 and RP by 48.50.*

$MD\% = \frac{1}{5}$ or 0.20 — *Reduce fraction, or change to decimal.*

29. $RP = \frac{NP}{1 - MD\%}$ — *The given formula.*

$60 = \frac{NP}{1 - 0.05}$ — *Replace RP by 60 and MD% by 0.05.*

$60 = \frac{NP}{0.95}$ — *Subtract first.*

$57 = NP$ — *Multiply both sides by 0.95.*

Exercise Set 3-2 Formulas

33. $A = P(1 + R)^n$ *The given formula.*

$2645 = P(1 + 0.15)^2$ *Replace A by 2,645, R by 0.15, and n by 2.*

$2645 = P(1.15)^2$ *Add first.*

$2645 = P(1.3225)$ *$1.15^2 = (1.15)(1.15) = 1.3225$.*

$2000 = P$ *Divide both sides by 1.3225.*

37. $d = \frac{D}{N}$ *The given formula.*

$12.50 = \frac{D}{70{,}000}$ *Replace d by 12.50 and N by 70,000.*

$875{,}000 = D$ *Multiply both sides by 70,000.*

41. Area = (length)(width) *The formula in words.*

42 = (length)(6) *Replace area by 42 and width by 6.*

7 = length *The length is 7 feet.*

45. Interest = (Principal)(Rate)(Time) *The formula in words.*

125 = (Principal)(0.10)(5) *Replace I by 125, R by 0.10 and T by 5.*

125 = (Principal)(0.50) *Multiply first.*

250 = Principal *The Principal is $250.*

Exercise Set 3-3 Ratio and Proportion

1. The ratio of 2 to 15 is written $\frac{2}{15}$.

5. The ratio of 20 to 25 is written $\frac{20}{25} = \frac{4}{5}$.

9. The ratio of 10 gallons to 3 gallons is written $\frac{10}{3}$.

13. Since 1 hour is 60 minutes, the ratio is written $\frac{5}{60} = \frac{1}{12}$.

17. Since 5 tons is 10,000 pounds, the ratio is written $\frac{200}{10{,}000} = \frac{1}{50}$.

21. The ratio of 1 inch to 2.54 centimeters is written $\frac{\text{1 inch}}{\text{2.54 centimeters}}$.

25. The ratio of 100 miles to 3 gallons of gasoline is written $\frac{\text{100 miles}}{\text{3 gallons}}$.

29. The ratio of 8 correct answers to 15 questions asked is written $\frac{\text{8 correct}}{\text{15 total}}$

Exercise Set 3-3 Ratio and Proportion

33. The ratio of 45 bushels to 1 acre is written $\frac{45 \text{ bushels}}{1 \text{ acre}}$.

37. The ratio of 15 ounces of fat to 4 pounds body weight is written $\frac{15 \text{ ounces}}{4 \text{ pounds}}$.

41. $\frac{7{,}040 \text{ yards}}{4 \text{ miles}} = \frac{11{,}440 \text{ yards}}{6.5 \text{ miles}}$

45. $\frac{95 \text{ miles}}{4 \text{ gallons}} = \frac{237.5 \text{ miles}}{10 \text{ gallons}}$

49. $\frac{1 \text{ inch}}{77 \text{ miles}} = \frac{6\frac{3}{4} \text{ inches}}{520 \text{ miles}}$

53. $\frac{4}{y} = \frac{32}{56}$ *The given proportion.*

$224 = 32y$ *The means-extremes product property.*

$7 = y$ *Divide both sides by 32.*

57. $\frac{2}{3} = \frac{5}{b}$ *The given proportion.*

$2b = 15$ *The means-extremes product property.*

$b = \frac{15}{2}$, or $7\frac{1}{2}$ *Divide both sides by 2.*

61. $\frac{y}{1.2} = \frac{80}{4.8}$ *The given proportion.*

$4.8y = 96.0$ *The means-extremes product property.*

$y = 20$ *Divide both sides by 4.8.*

65. $\frac{0.06}{b} = \frac{20}{150}$ *The given proportion.*

$9 = 20b$ *The means-extremes product property.*

$0.45 = b$ *Divide both sides by 20.*

Exercise Set 3-4 Applications

1. Let x = the number of hours worked

$3.75x = 123.75$ *(Wage rate)(Hours worked) = Gross earnings*

$x = 33$ *Divide both sides by 3.75.*

Thus, Kerry worked 33 hours last week.

5. Let w = the weight of the contents

$w + 9 = 84$ *Box + Contents = Total weight*

$w = 75$ *Subtract 9 from both sides.*

The weight of the contents is 75 pounds.

Exercise Set 3-4 Applications

9. Let a = the amount of sales over $8,000

$0.06(8000) + 0.075a = 988.50$

$480 + 0.075a = 988.50$

$0.075a = 508.50$

$a = 6780$

Thus, the dollar amount on sales over $8000 is $6780.

13. Let x = the sales tax on $26.00.

$\frac{0.09}{2.00} = \frac{x}{26.00}$ *tax/purchase = tax/purchase*

$2.34 = 2.00x$ *Cross-multiply.*

$1.17 = x$ *Divide both sides by 2.00.*

The sales tax is $1.17.

17. Let d = the distance traveled on 15 gallons.

$\frac{196}{7} = \frac{d}{15}$ *miles/gallons = miles/gallons*

$2940 = 7d$ *Cross-multiply.*

$420 = d$ *Divide both sides by 7.*

The car can go about 420 miles.

21. Let a = the amount of insurance.

$\frac{4.90}{1000} = \frac{78.40}{a}$ *premium/policy = premium/policy*

$4.90a = 78,400$ *Cross-multiply.*

$a = 16,000$ *Divide both sides by 4.90.*

Gladys bought $16,000 in insurance.

25. Let c = the carrying charge on $1500.

$\frac{9.80}{600} = \frac{c}{1500}$ *charge/balance = charge/balance*

$14,700 = 600c$ *Cross-multiply.*

$24.5 = c$ *Divide both sides by 600.*

The carrying charge would be $24.50.

Exercise Set 3-4 Applications

29. Let m = the number of miles in 10,000 yards.

$$\frac{1}{1760} = \frac{m}{10{,}000} \qquad \frac{\textit{miles}}{\textit{yards}} = \frac{\textit{miles}}{\textit{yards}}$$

$10{,}000 = 1760\ m$ *Cross-multiply.*

$5.68 = m$ *Divide both sides by 1760.*

To the nearest tenth, 10,000 yards = 5.7 miles.

PART C Chapter Three Sample Test

In 1-3, determine whether or not the given value of the variable is a solution of the stated equation.

1. $5x + 1 = 21$ and $x = 4$ — 1. ________

2. $2y + 3 = 3y - 7$ and $y = 9$ — 2. ________

3. $3(y + 2) - 5 = 2y + 11$ and $y = 10$ — 3. ________

In 4-8, solve and check each equation.

4. $8x - 9 = 39$ — 4. ________

5. $7y = 3y + 40$ — 5. ________

6. $5z + 1 = 3z + 19$ — 6. ________

7. $\frac{2}{3}t = 5 - t$ — 7. ________

8. $\frac{1}{2}k + 7 = 22 - \frac{1}{2}k$ — 8. ________

In 9-11, evaluate each formula for the given values.

9. $A = 2l + 2w$ for $l = 12$ and $w = 7$. — 9. ________

10. $F = \frac{9}{5}C + 32$ for $C = 70$ — 10. ________

11. $MS\% = \frac{M}{SP}$ for $MS\% = 0.3$ and $SP = 15$. — 11. ________

In 12-14, write as a ratio using the same units.

12. 9 inches to 5 feet — 12. ________

13. 6 cents to one dollar — 13. ________

14. 10 minutes to one hour — 14. ________

In 15 and 16, write each ratio with different units. Write the units in the answer.

15. 2 gallons of paint covers 625 square feet — 15. ________

16. 3 hours of work for $20.25 pay — 16. ________

In 17 and 18, solve each proportion.

17. $\frac{x}{12} = \frac{70}{168}$ — 17. ________

18. $\frac{48}{4} = \frac{9}{y}$ — 18. ________

In 19 and 20, solve each applied problem using a linear equation in one variable or a proportion.

19. Kim Lee worked 40 hours last week at $6.00 an hour, and some overtime at $9.00 an hour. If her gross earnings for the week were $271.50, how many hours overtime did Kim work last week? — 19. ________

20. If Jack Twilly can buy $2,500 of one year of term life insurance for $12, then at that rate how much would he pay for $30,000 of one year of term life insurance? — 20. ________

CHAPTER FOUR

PERCENTAGES

PART A Summary of Topics

OBJECTIVES

1. *Learn the meaning of percent.*
2. *Change a decimal fraction to a percent.*
3. *Change a common fraction to a percent.*
4. *Compute a percentage as a base times a rate.*
5. *Compute a sales tax as a percentage of an item's selling price.*
6. *Compute an interest as a percentage of a principal amount.*
7. *Compute a discount as a percentage of an item's cost.*
8. *Compute a base as a percentage divided by a rate.*
9. *Compute the price of an item when the sales tax and tax rate are known.*
10. *Compute the principal when the interest and interest rate are known.*
11. *Find the regular price of an item when the discount and discount rate are known.*
12. *Compute a rate as a percentage divided by a base.*
13. *Find the percent of sales tax when the sale price and sales tax are known.*
14. *Find the percent of interest when the principal and interest are known.*
15. *Find the percent of discount when the sale price and amount of discount are known.*

SECTION 4-1 Common Fractions, Decimal Fractions, and Percents

1. The word percent means parts per one hundred.

   As a consequence, the percent symbol can always be replaced with a fraction bar and a denominator of 100.

   $5\% = \frac{5}{100}$ $\qquad 112\% = \frac{112}{100}$ $\qquad 0.3\% = \frac{0.3}{100}$

2. The following two steps can be used as a procedure for writing a decimal fraction as a percent.

   Step 1. Move the decimal point in the fraction two places to the right.

   Step 2. Write a percent symbol (%) to the right of the number.

   To change a percent to a decimal fraction, the two steps are reversed. That is,

   Step 1. Drop the percent symbol.

   Step 2. Move the decimal point two places to the left.

3. To write a common fraction as a percent, change the common fraction to a decimal fraction, then use the two-step procedure described above.

   If the common fraction converts to an exact decimal fraction, (that is, the division yields a zero remainder), then the conversion process is straight-forward.

SECTION 4-1 Common Fractions, Decimal Fractions, and Percents

$\frac{1}{2} = 0.5 = 50\%$

$\frac{3}{4} = 0.75 = 75\%$

$\frac{1}{8} = 0.125 = 12.5\%$

If the common fraction does not convert to an exact decimal, then the percent may include a common fraction part, or the percent may be rounded to approximate the common fraction.

$\frac{1}{3} = 33\frac{1}{3}\%$ *(Exact)* $\qquad$ $\frac{1}{3} = 33.3\%$ *(Approximate)*

$\frac{5}{6} = 83\frac{1}{3}\%$ *(Exact)* $\qquad$ $\frac{5}{6} = 83.33\%$ *(Approximate)*

4. Percents such as $3\frac{5}{9}\%$ and $12\frac{2}{3}\%$ are usually changed to common fractions when the percents are used in a computation. The following two-step procedure can be used to write such a percent as a common fraction.

   Step 1. Write the mixed number as an improper fraction.

   Step 2. Drop the percent symbol and multiply the denominator of the fraction in Step 1 by 100. Reduce the fraction, if possible.

$$3\frac{5}{9}\% = \frac{32}{9}\% = \frac{32}{9} \cdot \frac{1}{100} = \frac{32}{900} = \frac{8}{225}$$

$$12\frac{2}{3}\% = \frac{38}{3}\% = \frac{38}{3} \cdot \frac{1}{100} = \frac{38}{300} = \frac{19}{150}$$

SECTION 4-2 Finding the Percentage

1. A percentage is a fractional part of a whole.

   Suppose that the whole is $100. Then $25 is a fractional part, or percentage of $100. This fact can be written in three equivalent ways:

   $25 is $\frac{1}{4}$ of $100 *(A common fraction)*

   $25 is 0.25 of $100 *(A decimal fraction)*

   $25 is 25% of $100 *(A percent)*

   In this illustration, $100 is the Base, 25% is the Rate, and $25 is the Percentage.

2. If P stands for percentage, or part,

   B stands for base

   R stands for rate

   then in any problem involving percentage, base and rate:

SECTION 4-2 Finding the Percentage

$$P = B \times R$$

In words: Percentage = Base times Rate

3. The amount of sales tax levied is usually a percent times the selling price of an item. Thus,

   P = the amount of the sales tax.

   B = the selling price of the item.

   R = the rate at which the tax is levied.

4. The amount of simple interest earned in one year by an amount of money on deposit is usually a percent times the amount on deposit. Similarly, the amount of simple interest charged in one year for borrowing an amount of money is usually a percent times the amount borrowed. Thus,

   P = the amount of interest earned (or charged).

   B = the amount of money on deposit (or borrowed).

   R = the rate at which interest is earned (or charged).

   The time for which money is deposited or borrowed is also a factor that effects the amount of interest. However, in this section the time is considered to be one year.

5. The amount of discount deducted from the selling price of an item is frequently a percent times the selling price of the item. Thus,

   P = the amount of the discount.

   B = the selling price of the item.

   R = the rate at which the discount is calculated.

SECTION 4-3 Finding the Base

1. If a Percentage (P) and Rate (R) are known, then the Base (B) can be calculated with the following equation:

$$B = \frac{P}{R}$$

   "The Base equals the Percentage divided by the Rate."

2. In a sales tax and selling price problem, the sales tax is the percentage and the selling price is the base. Thus

$$\text{Selling price} = \frac{\text{Amount of sales tax}}{\text{Rate of sales tax}}$$

3. In an interest and principal (amount of money) problem, the interest is the percentage and the principal is the base.

$$\text{Principal} = \frac{\text{Interest}}{\text{Rate of interest}}$$

SECTION 4-3 Finding the Base

4. In a discount and regular price problem, the discount is the percentage and the regular price is the base.

$$\text{Regular price} = \frac{\text{Discount}}{\text{Rate of discount}}$$

SECTION 4-4 Finding the Rate

1. If a Percentage (P) and Base (B) are known, then the Rate (R) can be calculated with the following equation:

$$R = \frac{P}{B}$$

"The Rate equals the Percentage divided by the Base."

2. In a sales tax and selling price problem, the sales tax is the percentage and the selling price is the base. Thus,

$$\text{Rate of tax} = \frac{\text{Amount of sales tax}}{\text{Selling price}}$$

3. In an interest and principal (amount of money) problem, the interest is the percentage and the principal is the base.

$$\text{Rate of interest} = \frac{\text{Interest}}{\text{Principal}}$$

4. In a discount and regular price problem, the discount is the percentage and the regular price is the base.

$$\text{Rate of discount} = \frac{\text{Discount}}{\text{Regular price}}$$

PART B Selected Solutions

Exercise Set 4-1 Common Fractions, Decimal Fractions, and Percents

1. 0.33 = 33% — *Move decimal point two places to the right and insert the % symbol.*

5. 0.076 = 7.6%

9. 4.2 = 420%

13. 65% = 0.65 — *Drop the % symbol and move the decimal point two places to the left.*

17. 0.75% = 0.0075

21. $\frac{1}{2} = 0.50 = 50\%$ — *Change $\frac{1}{2}$ to a decimal fraction, then change the decimal fraction to a percent.*

25. $\frac{5}{2} = 2.50 = 250\%$

29. $\frac{17}{20} = 0.85 = 85\%$

33. $\frac{1}{3} = 0.33\frac{1}{3}$ — *Divide to two decimal places and write remainder as a common fraction. Then change to a percent.*

$= 33\frac{1}{3}\%$

Exercise Set 4-1 Common Fractions, Decimal Fractions, and Percents

37. $\frac{5}{12} = 0.41\frac{2}{3}$

$= 41\frac{2}{3}\%$

41. $\frac{7}{30} = 0.23\frac{1}{3}$

$= 23\frac{1}{3}\%$

45. $3\frac{1}{3}\% = \frac{10}{3}\%$ — *Write the mixed number $3\frac{1}{3}$ as $\frac{10}{3}$.*

$= \frac{10}{300}$ — *Drop percent symbol and multiply 3 by 100.*

$= \frac{1}{30}$ — *Reduce the fraction.*

49. $2\frac{1}{7}\% = \frac{15}{7}\%$

$= \frac{15}{700}$

$= \frac{3}{140}$

53. $8\frac{2}{3}\% = \frac{26}{3}\%$

$= \frac{26}{300}$

$= \frac{13}{150}$

57. $\frac{2}{5} = 0.40$ (Decimal Fraction)

$= 40\%$ (Percent)

61. $75\% = 0.75$ (Decimal Fraction)

$= \frac{75}{100} = \frac{3}{4}$ (Common Fraction)

Exercise Set 4-2 Finding the Percentage

1. $P = 25\%$ of 1,200 — *$P = R \times B$*

$= 0.25(1{,}200)$ — *Write 25% as 0.25.*

$= 300$ — *Multiply.*

5. $P = 12\%$ of 37.5

$= 0.12(37.5)$

$= 4.5$

9. $P = 10.3\%$ of 2760

$= 0.103(2760)$

$= 284.28$

13. $P = 15\frac{3}{5}\%$ of 140

$= \frac{78}{500} \cdot 140$

$= \frac{546}{25}$ or 21.84

17. $P = 2\frac{2}{3}\%$ of 9.6

$= \frac{8}{300} \cdot 9.6$

$= \frac{6.4}{25}$ or 0.256

Exercise Set 4-2 Finding the Percentage

21. $P = 4\%$ of \$250.00   *P = R x B*

$= 0.04(\$250.00)$ *4% = 0.04*

$= \$10.00$ *Multiply.*

25. $P = 3\frac{1}{2}\%$ of \$560.00

$= 0.035(\$560.00)$

$= \$19.60$

29. $P = 4\frac{1}{3}\%$ of \$965.10

$= \frac{13}{300}(\$965.10)$

$= \$41.82$, to the nearest cent

33. Sales tax = Selling price x tax rate

Sales tax = \$40.00(0.065)   *$6\frac{1}{2}\% = 0.065$*

= \$2.60

37. Interest = Principal x rate of interest

Interest = \$12,400.00(0.129)   *12.9% = 0.129*

= \$1,599.60

41. Discount = Regular price x rate of discount

Discount = \$32.65(0.20)   *20% = 0.20*

= \$6.53

45. Number of women = Number of workers x Percent that are women

Number of women = 3,680(0.42)   *42% = 0.42*

= 1,546

49. Earnings from stock = Annual income x rate of income from stock

Earnings from stock = \$48,932(0.125)   *$12\frac{1}{2}\% = 0.125$*

= \$6,116.50

Exercise Set 4-3 Finding the Base

1. $\text{Base} = \frac{150}{5\%}$ $B = \frac{P}{R}$

$= \frac{150}{0.05}$ *5% = 0.05*

$= 3{,}000$ *Divide*

Exercise Set 4-3 Finding the Base

5. $\text{Base} = \frac{1,842}{0.6\%}$

$= \frac{1,842}{0.006}$

$= 307,000$

9. $\text{Base} = \frac{1783.5}{12.3\%}$

$= \frac{1783.5}{0.123}$

$= 14,500$

13. $\text{The number} = \frac{126.28}{8.2\%}$ *126.48 is the number and 8.2% is the rate.*

$= \frac{126.28}{0.082}$ *8.2% = 0.082.*

$= 1,540$ *Divide.*

17. $\text{The number} = \frac{11.76}{15\%}$ *11.76 is the percentage and 15% is the rate.*

$= \frac{11.75}{0.15}$ *15% = 0.15*

$= 78.4$ *Divide.*

21. $\text{Selling price} = \frac{\$0.74}{5\%}$ *Selling price* $= \frac{\textit{Sales tax}}{\textit{Rate of tax}}$

$= \frac{\$0.74}{0.05}$ *5% = 0.05*

$= \$14.80$ *Divide.*

The selling price of the level is $14.80.

25. $\text{Selling price} = \frac{\$2.49}{5\%}$ *Selling price* $= \frac{\textit{Sales tax}}{\textit{Rate of tax}}$

$= \frac{\$2.49}{0.05}$ *5% = 0.05*

$= \$49.80$ *Divide.*

The selling price of the bench vise is $49.80.

29. $\text{Value of Stocks} = \frac{\$724.58}{15.2\%}$ *Value of Stocks* $= \frac{\textit{Earnings}}{\textit{Rate of return}}$

$= \frac{\$724.58}{0.152}$ *15.2% = 0.152*

$= \$4766.97$ *Rounded to the nearest cent.*

The value of the stocks is about $4766.97.

33. $\text{Regular price} = \frac{\$31.50}{35\%}$ *Regular price* $= \frac{\textit{Discount}}{\textit{Rate of discount}}$

$= \frac{\$31.50}{0.35}$ *35% = 0.35*

$= \$90.00$ *Divide.*

The regular price of the dress was $90.00.

Exercise Set 4-3 Finding the Base

37. Number of students $= \frac{5,304}{34\%}$ — *Number of students* $= \frac{\textit{Students with jobs}}{\textit{Rate of job holders}}$

$= \frac{5,304}{0.34}$ — *34% = 0.34*

$= 15,600$ — *Divide.*

About 15,600 students are attending this college.

Exercise Set 4-4 Finding the Rate

1. Rate $= \frac{48}{160}$ — *Rate* $= \frac{\textit{Percentage}}{\textit{Base}}$

$= 0.3$ — *Divide.*

$= 30\%$ — *Write as a percent.*

5. Rate $= \frac{1107}{8200}$ — *Rate* $= \frac{\textit{Percentage}}{\textit{Base}}$

$= 0.135$ — *Divide.*

$= 13.5\%$ — *Write as a percent.*

9. Rate $= \frac{14,500}{174,000}$ — *Rate* $= \frac{\textit{Percentage}}{\textit{Base}}$

$= 0.08\frac{1}{3}$ — *Divide to two decimal places.*

$= 8\frac{1}{3}\%$ — *Write as a percent.*

13. Rate $= \frac{0.49}{9.8}$ — *Rate* $= \frac{\textit{Percentage}}{\textit{Base}}$

$= 0.05$ — *Divide.*

$= 5\%$ — *Write as a percent.*

17. Rate $= \frac{60}{500}$ — *60 is the Percentage, and 500 is the Base.*

$= 0.12$ — *Divide.*

$= 12\%$ — *Write as a percent.*

21. Rate $= \frac{0.51}{10.2}$ — *0.51 is the Percentage, and 10.2 is the Base.*

$= 0.05$ — *Divide.*

$= 5\%$ — *Write as a percent.*

Exercise Set 4-4 Finding the Base

25. $\text{Rate} = \frac{0.69}{4.6}$ — *0.69 is the Percentage, and 4.6 is the Base.*

$= 0.15$ — *Divide.*

$= 15\%$ — *Write as a percent.*

29. $\text{Rate of tax} = \frac{\$2.47}{\$95.00}$ — *The tax is the Percentage, and the selling price is the Base.*

$= 0.026$ — *Divide.*

$= 2.6\%$ — *Write as a percent.*

The sales tax rate is 2.6%.

33. $\text{Rate of interest} = \frac{\$1,725}{\$12,500}$ — *The interest is the Percentage, and the amount borrowed is the Base.*

$= 0.138$ — *Divide.*

$= 13.8\%$ — *Write as a percent.*

The interest rate is 13.8%

37. $\text{Rate of discount} = \frac{\$3.00}{\$12.50}$ — *The discount is the Percentage, and the regular price is the Base.*

$= 0.24$ — *Divide.*

$= 24\%$ — *Write as a percent.*

The discount rate is 24%.

41. Decrease in number unemployed = 460,000 - 381,000

= 79,000

$\text{Rate of decrease} = \frac{79,000}{460,000}$ — *The decrease in unemployed is the Percentage, and last month's figure is the Base.*

$= 0.1717$ — *Divide to four decimal places.*

$= 17.2\%$ — *To the nearest tenth of a percent.*

45. Number of units increase = 901 - 850

= 51 units

$\text{Rate of increase} = \frac{51}{850}$ — *Number of units increase is the Percentage, and number of units sold last month is the Base.*

$= 0.06$ — *Divide.*

$= 6\%$ — *Write as a percent.*

PART C Chapter Four Sample Test

In 1-4, write each decimal fraction as a percent.

1. 0.45   1. ____________
2. 2.3   2. ____________
3. 0.189   3. ____________
4. 0.035   4. ____________

In 5-8, write each percent as a decimal fraction.

5. 10.9%   5. ____________
6. 175%   6. ____________
7. 12%   7. ____________
8. 0.9%   8. ____________

In 9-12, write each percent as a reduced common fraction.

9. 20%   9. ____________
10. $3\frac{1}{3}\%$ 10. ____________
11. 160%   11. ____________
12. 7.5%   12. ____________

In 13-20, write each number in the other forms indicated.

Common Fraction	*Decimal Fraction*	*Percent*
$\frac{3}{8}$	13. ____________	14. ____________
15. ____________	0.15	16. ____________
17. ____________	18. ____________	22.5%
$\frac{11}{5}$	19. ____________	20. ____________

In 21-24, find the indicated percentages.

21. $7\frac{1}{2}$% of 1,500 — 21. ________

22. 12.3% of 62,000 — 22. ________

23. 6% of 15.5 — 23. ________

24. 0.8% of 635 — 24. ________

In 25-28, find the unknown base or the unknown rate.

25. The percentage is 252 and the rate is 10.5% — 25. ________

26. The percentage is 235 and the base is 940. — 26. ________

27. The rate is $3\frac{1}{3}$% and the percentage is 40. — 27. ________

28. The base is 15.25 and the percentage is 1.22. — 28. ________

29. Morrow's Mens Shop is having a sale on suits. A suit that regularly sells for $230.00 is on sale for $161.00. Find the rate of discount on this suit based on the regular price of the suit. — 29. ________

30. Ralph Jefferson bought a microwave oven at Holman TV and Appliance store. The sales tax on the oven was $15.30 and the sales tax rate was 4.5%. How much did Ralph pay for the oven? — 30. ________

CHAPTER FIVE

PAYROLL

PART A Summary of Topics

OBJECTIVES

1. *Know what is meant by gross earnings.*
2. *Compute gross earnings using a salary method.*
3. *Compute gross earnings using an hourly method.*
4. *Compute gross earnings using a unit work method.*
5. *Compute gross earnings using a commission method.*
6. *Know what is meant by a tax deduction.*
7. *Compute an individual's FICA tax deduction.*
8. *Compute an individual's Federal Income Tax (FIT) deduction.*
9. *Compute an individual's State Income Tax (SIT) deduction.*
10. *Compute an individual's state disability insurance (SDI) tax deduction.*
11. *Compute an employer's FICA tax.*
12. *Compute an employer's Federal Unemployment Tax Act (FUTA) tax.*
13. *Compute an employer's State Unemployment Tax Act (SUTA) tax.*
14. *Learn what is meant by voluntary deductions.*
15. *Learn what is meant by benefit income.*
16. *Learn what is meant by net pay.*
17. *Learn about payroll reports.*

SECTION 5-1 Gross Earnings

1. Gross earnings is the total amount of money earned, before any deductions, by an individual for service to a company or business.

2. Deductions are amounts of money subtracted from gross earnings.

   There are several reasons for deductions, or withholdings, from an individual's gross earnings. Income tax (both federal and state), savings, loan payments and union dues are among the most obvious types of deductions.

3. The hourly method is one way of computing an individual's gross earnings. Using this method:

   Gross earnings = (number of hours worked)(rate per hour)

   A time card is frequently used to keep a record of the number of hours worked for an hourly paid employee. Overtime pay is paid when an individual works more than what is considered regular for the individual's position.

4. The salary method is another way of computing an individual's gross earnings. Using this method:

   Gross earnings = a unit of pay for a specified period of time worked

   Usually a record is not kept of time spent on the job for individuals that are paid by the salary method.

5. The commission method is a third way of computing an individual's gross earnings. Using this method:

   Gross earnings = (amount of sales)(rate of commission)

The commission method is frequently used for individuals in sales, as a means of motivating the individual to increase the amount of sales. There are variations in this method that are worth noting. First, a base salary may be given by which an individual gets a daily, or weekly, or monthly salary. Then the commission is added to this base salary. Secondly, the rate of commission may be increased for higher quotas on sales. Such higher commission rates provide further incentives to an individual to increase the dollar amounts in sales.

6. The unit-work method is a fourth way of computing an individual's gross earnings. Using this method:

   Gross earnings = (number of units)(pay per unit)

   The term piece work is sometimes used for jobs that are based on the unit-work method. This method is frequently used to motivate an individual to achieve higher levels of productivity by turning out more units in the same period of time.

SECTION 5-2 Employee Tax Deductions

1. A tax deduction can be described as a dollar amount that is computed based on gross earnings. Such dollar amounts are subtracted from an individual's gross earnings by the employer and forwarded to the government for whom the tax was collected.

   The social security tax is one such tax deduction. Other examples include federal income tax (FIT), state income tax (SIT), and in some cities a city income tax.

2. The Federal Insurance Contributions Act (FICA) tax is used by the federal government to finance the Social Security Fund.

   The following features of this tax are significant:

   a. Both employee and employer pay a FICA tax. Although the tax rate is usually the same for both, the rate can differ, as it did in 1984.

   b. The tax is paid on gross earnings up to some specified wage limit. The wage limit is levied by law and has been gradually increasing over the past few years.

   c. The wage limit is based on accumulated earnings that begin on January 1 of each year, and includes all earnings up to, but not including, the present pay period.

3. The Federal Income Tax (FIT) is one that is frequently deducted from an individual's gross earnings. To calculate the amount of Federal Income Tax to deduct from an individual's gross earnings, four items of information are needed:

   1. The marital status of the individual.
   2. The number of exemptions claimed by the individual.
   3. The amount of the gross earnings.
   4. The pay period on which the earnings are computed.

   The tax can be calculated with a formula. However, the usual method is to use tax tables that are provided by the Internal Revenue Service (IRS).

4. A State Income Tax (SIT) is another tax that is frequently deducted from an individual's gross earnings:

The SIT may be computed in a manner that is similar to the federal income tax. For such states, tables are provided for computing the amount of a deduction. In some states a flat tax rate is used and the rate is applied to all taxable income to a specified wage limit.

5. A State Disability Insurance (SDI) tax may also be deducted from an individual's gross earnings. This tax is used to finance payments to workers who are injured or disabled as a result of a job-related accident.

The tax is usually a flat-tax that has a wage-limit, both of which are controlled by the state government in which the tax is collected.

SECTION 5-3 Employer Payroll Taxes

1. The employer must also pay a FICA tax for each employee that must pay the tax.

In the past, the wage limit and tax rate of employee and employer have been the same. However, in 1984 and 1985, the limits and rate for employee and employer were different. For this text, the employer's tax rate is set at 7%, and the wage limit is $37,800.

2. An employer must also contribute toward a Federal Unemployment Tax Act (FUTA) program. These funds are used to finance a federally funded unemployment program.

In this text a $7,000 wage limit and a 0.8% tax rate will be used for the FUTA tax.

3. An employer must also contribute toward a State Unemployment Tax Act (SUTA) program. These funds are used to finance a state unemployment program.

In this text a $7,000 wage limit and a 2.7% tax rate will be used for the SUTA tax.

SECTION 5-4 Voluntary Deductions and Benefit Income

1. A voluntary deduction is one that is authorized by an employee for personal reasons.

Some of the more common reasons for voluntary deductions are group insurance payments, union (or association) dues, loan payments, pension contributions, payments to a vacation fund, and deposits in a savings plan.

2. Benefit income is earnings that an employee may get on a job that is in addition to the regular gross earnings from the employer.

Some of the more common sources of benefit income are tips, paid insurance (such as health, dental and life), vacation pay and bonuses.

3. Since benefit income may be subject to FICA, FIT, and SIT taxes, the term taxable gross income will be used.

Taxable gross income = regular gross earnings + benefit income

Some benefit income may not be subject to such taxes. For example, many employers pay the premiums on health and dental programs for employees. Such payments are benefit income, but the payments may not be added to regular gross income and taxed.

SECTION 5-5 Net Pay and Payroll Reports

1. An individual's net pay is the difference between the gross earnings for a pay period and the total deductions. Written as an equation:

   Net pay = Gross earnings - Total deductions

   A pay stub is frequently attached to an individual's pay check that provides a written account of the gross earnings, the itemized deductions, and the resulting net pay.

2. A business must also maintain a payroll journal for each payroll period. Such information as gross earnings, tax deductions, voluntary deductions, and net pays of all employees for the pay period are recorded.

3. An individual payroll record must be maintained for each individual employee of a given business. Such a record shows a running-total for each of the items shown on a payroll journal.

   As each item from a payroll journal is entered on the individual payroll record, the accumulated totals should be checked to see whether or not any of the wage limits for FICA, FUTA, SUTA, and SDI have been reached. Such information is needed when making out subsequent payrolls on the employees.

PART B Selected Solutions

Exercise Set 5-1 Gross Earnings

1. *Gross earnings = (number of hours worked)(rate per hour)*

   Gross earnings = (40)(8.90)

   = 356.00

   Thus, the gross earnings of James Koeth are $356.00.

5. Gross earnings = (40)(12.40) + (5)(12.40)(1.5) *1.5 times regular rate for overtime.*

   = 496.00 + 93.00

   = 589.00

   Thus, the gross earnings of Susan Riddle are $589.00.

9. On the time card in Figure 5-A:

   40 hours regular time at $6.80 per hour

   4 hours overtime at $13.60 per hour

   Gross earnings = (40)(6.80) + (4)(13.60)

   = 272.00 + 54.40

   = 326.40

   Thus, the gross earnings of Bill Board are $326.40.

13. On the time card in Figure 5-E:

40 hours regular time at \$9.10 per hour

2.5 hours overtime at \$13.65 per hour

$$\text{Gross earnings} = (40)(9.10) + (2.5)(13.65)$$
$$= 364.00 + 34.125$$
$$= 398.125$$

Thus, the gross earnings of Emmanuel Washington are \$398.13.

17. $$\text{Monthly salary} = \frac{\text{annual salary}}{12}$$
$$= \frac{18{,}600}{12}$$
$$= 1550$$

Thus, the gross monthly earnings of Ron Bronikowski are \$1550.00.

21. *Gross earnings = (amount of sales)(rate of commission)*

$$\text{Gross earnings} = (10{,}450)(0.0225) \qquad 2.25\% = 0.0225$$
$$= 235.125$$

Thus, the gross earnings of Barney Hines are \$235.13.

25. *Gross earnings = Base salary + commission*

$$\text{Gross earnings} = 400.00 + (4{,}836)(0.04)$$
$$= 400.00 + 193.44$$
$$= 593.44$$

Thus, the gross earnings of Susan Hadley are \$593.44.

29. *Gross earnings = (number of units)(pay per unit)*

$$\text{Gross earnings} = (1{,}500)(0.35)$$
$$= 525.00$$

Thus, the gross earnings of Sharon Small are \$525.00.

33.

Wood doors:	(3)(25.00) =	75.00
Wood doors with glass windows:	(5)(35.00) =	175.00
Metal doors:	(2)(50.00) =	100.00
	Total	350.00

Thus, the gross earnings of Sam Cloud are \$350.00.

Exercise Set 5-2 Employee Tax Deductions

1. Write \$215.10 as \$200.00 + \$15.10

   The tax from the table is \$13.40 + \$1.01 = \$14.41

   The tax on \$200 is located in the lower right corner of the right page of Table B-2. The tax on \$15.10 is found on line 12 to the right of 15.00 - 15.15 under the column headed "Tax to be withheld".

5. Write \$388.40 as \$300.00 + \$88.40

   The tax from the table is \$20.10 + \$5.92 = \$26.02

   Thus the FICA tax withheld is \$26.02.

9. FICA tax = (1,206.03)(0.067)

   = 80.80 — *Round the tax to the nearest cent.*

   The FICA tax withheld is \$80.80.

13. Since Ken Baker claims single status, use Table B-7 . Locate 440 under the column headed "At least" and 450 under the column headed "But less than". To the right under the column headed "And the number of withholding allowances claimed is 0", read 82.80. Thus the FIT deduction is \$82.80.

17. Since Edie Wong claims married status, use Table B-5 . Locate 430 and 440 in the left columns. Under the column headed 0, to the right of 430 and 440, read 62.50. Thus the FIT deduction is \$62.50.

21. Since Fred Young claims married status, use Table B-5. Locate 580 and 590 in the left columns. Under the column headed 5, to the right of 580 and 590, read 75.40. Thus the FIT deduction is \$75.40.

25. Since Nancy Wolfe claims married status, use Table B-8 . Locate 640 and 660 in the left columns. Under the column headed 2, to the right of 640 and 660, read 19.50. Thus the SIT deduction is \$19.50.

29. The state tax = (821.60)(0.015) — *$1\frac{1}{2}\% = 0.015$*

   = 12.32 — *Round to two decimal places.*

   Thus the tax withheld is \$12.32.

33. SDI tax = (1,050.50)(0.009) — *0.9% = 0.009*

   = 9.45 — *Round to two decimal places.*

   The tax withheld is \$9.45.

37\. Wage limit subject to tax $21,000.00

Accumulated earnings 20,587.10

Current gross earnings subject to tax $ 412.90

SDI tax = (412.90)(0.009) *0.9% = 0.009*

= 3.72 *Round to two decimal places.*

The tax withheld is $3.72.

Exercise Set 5-3 Employer Payroll Taxes

1. The current gross earnings are subject to the three employer taxes.

   a. FICA tax = (427.00)(0.07) *7.0% = 0.07*

      = 29.89 *Round to two decimal places.*

      Employer's FICA tax is $29.89.

   b. FUTA tax = (427.00)(0.008) *0.8% = 0.008*

      = 3.42 *Round to two decimal places.*

      Employer's FUTA tax is $3.42.

   c. SUTA tax = (427.00)(0.027) *2.7% = 0.027*

      = 11.53 *Round to two decimal places.*

      Employer's SUTA tax is $11.53.

5. The current gross earnings are subject to the FICA tax.

   a. FICA tax = (1175.50)(0.07) *7.0% = 0.07*

      = 82.29 *Round to two decimal places.*

      The employer's FICA tax is $82.29.

   b. Wage limit $7,000.00

      Accumulated earnings 6,418.75

      Current earnings subject to FUTA and SUTA tax $ 581.25

      FUTA tax = (581.25)(0.008)

      = 4.65

      The FUTA tax is $4.65.

   c. SUTA tax = (581.25)(0.027)

      = 15.69

      The SUTA tax is $15.69.

9. a. Wage limit for FICA tax $37,800.00
Accumulated gross earnings 36,185.27
Current earnings subject to tax $ 1,614.73

FICA tax = (1614.73)(0.07)

= 113.03

The FICA tax is $113.03.

b. and c. The wage limit for FUTA and SUTA taxes is $7,000. Thus the current gross earnings are not subject to these taxes.

13. The current gross earnings are subject to FICA, FUTA and SUTA taxes.

a. FICA tax = (914.73)(0.07)

= 64.03

The FICA tax is $64.03.

b. FUTA tax = (914.73)(0.008)

= 7.32

The FUTA tax is $7.32.

c. SUTA tax = (914.73)(0.027)

= 24.70

The SUTA tax is $24.70

Exercise Set 5-4 Voluntary Deductions and Benefit Income

1. a. Employee's contribution = (845.50)(0.02)

= 16.95

Thus, Dennis Kipps contributes $16.95 to the retirement plan.

b. Employer's contribution = Employee's contribution

= $16.95

Total contribution = 2($16.95) = $33.90

5. Vacation pay deduction = (897.50)(0.025) *$2\frac{1}{2}\% = 0.025$*

= 22.44 *Rounded to two decimal places.*

The vacation pay deduction is $22.44.

9. Pension plan deduction = (1,377.80)(0.015)

= 20.67

Deduction		Amount
Pension plan		$ 20.67
Car payment		230.00
Union dues		19.50
	Total	$270.17

13. Regular gross earnings = $526.17
    Benefit income = 5.26 — *1% of gross earnings*
    Taxable income = $531.43

    a. FICA tax = (531.43)(0.067) — *6.7% = 0.067*

       = 35.61 — *Round to two decimal places*

       The FICA tax is $35.61.

    b. From Appendix Tables B-4 and B-5, with three dependents,

       FIT tax = $72.50

    c. From Appendix Tables B-8 and B-9, with three dependents,

       SIT tax = $11.80

Exercise Set 5-5 Net Pay and Payroll Reports

1. For the payroll record of S. Yushiko:

   Earnings:

(40)(8.75)	=	$350.00
(3)(13.20)	=	39.60
Total		$389.60

   Deductions:

FICA	$26.10
FIT	61.90
SDI	3.51
SIT	14.20
Savings	38.90
Total	$144.61

   Net pay = $389.60 - 144.61

   = $244.99

5. For the payroll record of Ed Hollis:

   a. Enter 32 hours regular time at $15.62 per hour = $499.84

   b. Enter 4 hours overtime at $23.43 per hour = 93.72

      Total gross earnings $593.56

   c. All the gross earnings are subject to FICA tax.

      (593.56)(0.067) = $ 39.77

   d. Using Appendix Tables B-6 and B-7, with two dependents:

      FIT = $117.20

   e. All the gross earnings are subject to SDI tax.

      (593.56)(0.009) = $ 5.34

f. Using Appendix Table B-10, with two dependents:

SIT = $ 35.19

g. Vacation pay deduction:

(593.56)(0.03) = 17.81

Total deductions $215.31

Net pay = $593.56 - 215.31 = $378.25

9. In the Employee's Individual Earnings Record in Figure 5-Q,

a. Enter 373.49 in the Total Earnings column

b. Enter 25.02 in the FICA column

c. Enter 64.20 in the FIT column

d. Enter 14.20 in the SIT column

e. Enter 3.36 in the SDI column

f. Enter 11.20 in the STOCK column

g. Enter 255.53 in the NET PAY column

h. Enter 8258.55 in the Cumulative Total column

Notice that the Cumulative Total is the sum of the previous Cumulative Total (7,885.06) plus the current Total Earnings (373.49).

PART C Chapter Five Sample Test

In 1-3, compute the gross earnings of each hourly paid employee.

	Employee	Regular Time	Overtime	Regular Rate	Overtime Rate		
1.	Randy Lee	40	$1\frac{1}{2}$	$ 7.30	$11.25	1.	________
2.	Joan Gault	32	4.3	$ 8.50	$12.75	2.	________
3.	Dave Molinari	40	0	$10.60	$21.20	3.	________

In 4-6, compute the gross earnings of each employee based on the given sales amounts and rates of commission.

	Employee	Amount of Sales	Rate of Commission		
4.	Travis Wong	$16,875.90	2.25%	4.	________
5.	Hector Sanchez	$77,940.00	$1\frac{3}{8}$ %	5.	________
6.	Gus Heath	$ 8,995.00	5.2%	6.	________

In exercises 7-9, compute the gross earnings of each employee for the week based on the given information.

7. Hilary Thomas receives a weekly salary of $215.00 as manager of Tuttle's Tea House. She also gets commission on all sales. Last week the sales were $5,827.90. What was Hilary's salary last week? 7. ________

8. Nick Triandos works at Upton Radial Tire Store. His base weekly salary is $180.00. He gets an additional $27.50 for each brake job he does. Last week Nick had 7 brake jobs. What was his weekly salary? 8. ________

9. Wanda Day makes phone calls for a newspaper to obtain new customers. She gets $0.35 for each phone call, and an additional $1.75 for each new subscriber. Last week Wanda made 482 calls and signed up 68 new subscriptions. What was her weekly salary? 9. ________

In 10-12, use a 6.7% rate and $37,800 wage limit to compute the FICA tax for each gross earnings.

	Employee	*Accumulated Earnings*	*Current Gross Earnings*	
10.	Le Pham	$12,987.50	$ 427.80	10. ______________
11.	Nestor King	$ 6,802.09	$1,738.50	11. ______________
12.	Thad Ewing	$36,845.20	$4,769.25	12. ______________

In 13-15, the employees of Digital Techtronics are paid based on the number of components assembled each week.

For units	*1-20*	*$6.80 each*
For units	*21-25*	*$7.25 each*
For units	*26-30*	*$7.90 each*
For units	*over 30*	*$8.50 each*

13. Lynn Gonzales assembled 23 units 13. ______________

14. Carli Symon assembled 18 units 14. ______________

15. Ester Washington assembled 27 units 15. ______________

In 16-18, for each gross earnings:

a. Use a 7.0% tax rate and a $37,800 wage limit to compute the employer FICA tax.

b. Use a 0.8% rate and a $7,000 wage limit to compute the employer FUTA tax.

c. Use a 2.7% rate and a $7,000 wage limit to compute the employer SUTA tax.

	Employee	*Accumulated Earnings*	*Current Gross Earnings*	
16.	Tracy Shephard	$3,870.50	$ 319.72	16. a. ______________
				b. ______________
				c. ______________
17.	Glenn Bonelli	$5,074.29	$ 816.75	17. a. ______________
				b. ______________
				c. ______________
18.	Freida Lang	$6,810.00	$1,450.00	18. a. ______________
				b. ______________
				c. ______________

CHAPTER SIX

THE MATHEMATICS OF BUYING

PART A Summary of Topics

OBJECTIVES

1. *Learn how to read an invoice.*
2. *Learn some of the abbreviations used on an invoice.*
3. *Learn how to extend an invoice.*
4. *Compute a volume discount based on case lots.*
5. *Compute a volume discount based on the dollar amount per order.*
6. *Compute an incremental volume discount based on the number of items purchased.*
7. *Compute an annual volume discount.*
8. *Compute a discount based on time.*
9. *Learn what is meant by the end-of-month method of dating.*
10. *Learn what is meant by the receipt-of-goods method of dating.*
11. *Know what is meant by extra dating.*
12. *Learn the definition of a trade discount.*
13. *Compute a single discount.*
14. *Compute the net cost after a discount.*
15. *Compute the discount rate when the discount is known.*
16. *Compute a series or chain discount.*

SECTION 6-1 Invoice

1. An invoice is a document prepared by the seller that identifies the goods or services and the terms of the sale.

   Most of the headings in an invoice are self-evident, such as Sold To, Invoice Date, Stock No, and so on. However, for some businesses it may be necessary to include some headings that are unique to that business. It may therefore be necessary to include a glossary with the invoice to identify the meaning of such headings.

2. Usually two tasks need to be performed to extend an invoice.

   Task 1. For each item on the invoice, use the equation

   Extended amount = number of units · cost per unit

   Task 2. Find the sum of the values in the Extension column to find the invoice total, before any taxes, shipping, or other costs or discounts are taken into account.

SECTION 6-2 Discounts Based on Volume

1. A discount is some type of a benefit to a buyer. Frequently the form of a benefit is a cash reduction in the cost of the product.

   Another form of a discount that differs from a cash reduction is the inclusion of one or more "free" items to the quantity ordered. For example, "buy three tires and get the fourth one free".

2. One type of volume discount is called a case discount.

   The number of items in a case depends in part on the quantity packed in so-called "case lots". If the items are boxed, then usually the case is simply the number of items in the box.

3. Another type of volume discount is based on the dollar amount of the order.

    Such a discount is frequently given when the order contains several different items with some common characteristic. For example, several different items that are all building supplies, or several different items in a grocery order for a food store.

4. Another type of volume discount is based on the number of items purchased over some specified period of time.

    This type of volume discount allows an individual or business to purchase small quantities over some period of time and earn a large volume discount. Notice that accurate records of such purchases must be maintained so as to be aware of discounts that may be earned by making a few additional purchases.

5. Another type of volume discount is based on the total volume of business for a year.

    This type of volume discount is similar to the one described in 4. The difference is that this discount may include a dollar amount. That is, if the dollar amount purchased over a specified time is $1,000,000, then a certain discount is earned. Thus, the volume is not dependent on the quantity purchased, but the dollar amount purchased.

SECTION 6-3 Discounts Based on Time

1. A discount may be offered to a buyer based on the time it takes the buyer to pay for the merchandise. Such a discount is called a time discount.

    There are two topics relevant to a time discount:

    1. the rate of discount and the period of time such a discount is available,

    2. the methods of determining when a time period begins.

2. The symbols N/P_1 and N/P_2 have the following meaning on the terms of an invoice:

    R represents the rate of discount for time period P_1.

    P_1 represents the number of days (period 1) in which the discount can be earned.

    N represents the net amount of the invoice.

    P_2 represents the number of days (period 2) in which the net amount can be paid.

    The paired notation R/P_1 N/P_2 can be extended to multiple rates and time periods. For example,

    $$R_1/P_1 \quad R_2/P_2 \quad N/P_3$$

    specifies a discount rate R_1 for a period P_1, then a discount rate R_2 for a period P_2, then the net amount for a period P_3.

3. The EOM (End-of-Month) method of dating starts a time period for determining discounts and net payments the first day of the month following the date on the invoice. The word prox can also be used to indicate this method of dating.

Some businesses will provide a little flexibility in the EOM method of dating. That is, when an order is delivered within a few days of the end of a given month, the time period starts not on the first of the next month, but on the first of the following month. Such a grace period is used to foster good business relationships.

4. The ROG (Receipt-of-Goods) method of dating starts a time period for determining discounts and net payments the first day after the goods are received by the buyer.

   With the ROG method of data, the first day of P_1 is the day after the goods are received by the buyer.

5. The symbol $R/P_1 - P_2$ Extra can be used to specify a second period of time P_2 during which a discount with rate R can be earned. The concept is called extra dating.

SECTION 6-4 Trade Discounts

1. A trade discount on an item is a discount given to a business or individual that is going to resell the item.

   Associated with a trade discount are the following terms:

   List - usually the suggested price for resale.

   Discount - the rate at which the discount is computed, and usually is based on the list price of the item.

   Net cost - the difference between the list price and discount.

2. Discount = list price · discount rate (Equation 1)

3. Net cost = list price - discount (Equation 2)

4. Discount rate = $\frac{\text{discount}}{\text{list price}}$ (Equation 3)

5. A series discount is one that consists of two or more discounts deducted from a list price. To illustrate,

   15/10 stands for a 15% discount, then a 10% discount on the difference obtained when the first discount has been deducted.

   On a series discount, the second, third, and subsequent discounts are based on the amount obtained after the previous discounts have been deducted. As a consequence, a 15/10 series discount is less than a 15 + 10 = 25% discount based on the given list price.

6. The net cost factor is a technique that can be used to change a series discount to an equivalent single discount. To illustrate, if R_1/R_2 is a series discount, then to find the net cost factor, form the product $(1 - R_1)(1 - R_2)$.

   Remember that the net cost factor should not be rounded before multiplying the list price by it to find the total discount specified by the series.

PART B Selected Solutions

Exercise Set 6-1 Invoices

1. There is 1 HP30 Heat Pump ordered.

5. The invoice shows that 3 Freon Connectors were shipped.

9. The description of the item on line 3 is Control Panel.

13. The number B-0 (stands for Back-Ordered) for item 3 is 1.

17. The shipping destination is Brian, Ohio.

21. The extension for line 1 is 1 · 3,385.00 = $3,385.00.

25.

Extension line 1	$3,385.00
Extension line 2	32.63
Extension line 3	98.70
Subtotal	$3,516.33
Shipping	31.60
Invoice Total	$3,547.93

Exercise Set 6-2 Discounts Based on Volume

1. Total purchase price = 10 cases · $12.00 per case

   = $120.00

   Discount = total purchase price · rate of discount

   = $120.00 · 0.10

   = $12.00

5. Total purchase price = 24 cases · $110.50 per case

   = $2,652.00

   Discount = total purchase price · rate of discount

   = $2,652.00 · 0.085

   = $225.42

9. Since $300 is less than $350, no discount is given on the order based on the dollar amount.

13. Since $2,550.00 is between $2,501 and $5,000, a 3% discount is given.

    Discount = $2,550.00 · 0.03

    = $76.50

17. Since 90 is less than 100, the order for January does not earn a volume discount.

Although there is no discount on this order, the 90 items processed this month will count in subsequent months to earn discounts on items processed later in the year.

21. Including the items in July:

90 + 80 + 140 + ... + 96 = 684 items were processed.

The cost of the items processed in July = 96 items • \$15 per item

= \$1440.

Since 684 is between 501 and 750, a 3% discount is earned.

Discount = \$1440 • 0.03

= \$43.20

25. Since 4,600 feet is less than 5,000 feet, Stein Electric will not get a discount based on an annual volume.

29. Since 6,200 is between 5,001 and 10,000, a 3% discount has been earned.

Total cost = 6,200 feet • \$0.76 per foot

= \$4,712.00

Discount = \$4,712.00 • 0.03

= \$141.36

33. a. Since 15 + 10 = 25, and 25 is less than 40, no discount is earned.

b. Furthermore, no additional discount is given.

37. a. Since 110 + 10 = 120, the items on this order will get a 3% discount. (Notice, 120 is in the 61-120 bracket on the discount schedule).

Total cost = 10 items • \$400 per item

= \$4000.00

Discount = \$4000.00 • 0.03

= \$120.00

b. Since the sum of the previous number of units purchased this year and the number of units on the last order of this year did not move into a new discount bracket, no additional discount is earned.

If the final order had been for 11 units and not 10, then 110 + 11 = 121, and 121 is in the 121-240 bracket. Thus, the current order of 11 would earn a 5% discount, and the previous 110 units would earn an additional 2% discount. The discount would have been

(11)(400)(0.05) + (110)(400)(0.02) = \$1100.00

The additional discount was more than enough to buy two units. This example emphasizes the need to maintain accurate accounting when annual volume discounts are involved.

Exercise Set 6-3 Discounts Based on Time

1. a. Discount = Subtotal • discount rate

= \$246.40 • 0.02 — *Rate of discount is 2%.*

= \$4.93 — *Amount of discount.*

b. Net amount = Subtotal - discount

= \$246.40 - 4.93

= \$241.47 — *If discount is earned.*

c. Net amount = \$246.40 — *The subtotal is due.*

5. a. Discount = \$246.40 • 0.035 — *Rate of discount is 3.5%.*

= \$8.62 — *Amount of discount.*

b. Net amount = \$246.40 - 8.62

= \$237.78 — *If discount is earned.*

c. Net amount = \$246.40 — *The subtotal is due.*

9. 2/15 means a 2% discount is earned if paid within 15 days.

N/45 means the net amount is due if paid between the 16th and the 45th days inclusive.

13. 1.5/30 means a 1.5% discount is earned if paid within 30 days.

N/60 means the net amount is due if paid between the 31st and the 60th days inclusive.

17. 4/10 means a 4% discount is earned if paid within 10 days.

2/30 means a 2% discount is earned if paid between the 11th and the 30th days inclusive.

N/45 means the net amount is due if paid between the 31st and the 45th days inclusive.

21. $2\frac{1}{3}$ /15 means a $2\frac{1}{3}$ % discount is earned if paid within 15 days.

$1\frac{1}{3}$ /30 means a $1\frac{1}{3}$ % discount is earned if paid between the 16th and the 30th days inclusive.

N/45 means the net amount is due if paid between the 31st and the 45th days inclusive.

25. a. With EOM dating, the discount period begins on the first day of the following month, namely 2-1. Since 3/15 is shown, the discount period has 15 days. Thus, the discount can be earned if the invoice is paid prior to 2/16.

b. Since N/45 is shown, the net amount period starts on 2/16 and continues for 30 days. (45 - 15 = 30). 2/16 through 2/28 is 13 days. 3/1 through 3/17 is 17 days. The net amount period is from 2/16 through 3/17.

c. A finance charge can be assessed on 3/18.

29. a. With ROG dating, the discount period begins on the day after the goods are received, (that is 12-11, since the goods were received on 12-10). With 4/10 showing, a 4% discount can be earned if the invoice is paid between 12-11 and 12-20 inclusive.

Discount = \$280.80 • 0.04

= \$11.23

b. Net amount = \$280.80 - 11.23

= \$269.57

c. With N/30 showing, the net amount is due if the invoice is paid within 30 days of the receipt of goods. Since 12/31 is within the alloted time, the net amount of \$280.80 must be paid.

33. a. With 3.5/15 showing, a 3.5% discount can be earned if the invoice is paid within the first 15 days of the receipt of goods, (that is 12-11 through 12-25).

Discount = \$280.80 • 0.035

= \$9.83

b. Net amount = \$280.80 - 9.83

= \$270.97

c. Since 12/31 is within the N/30 time period and outside the discount time period, the net amount of \$280.80 is due.

37. 3/20 - 20 Ex means that a 3% discount can be earned if the invoice is paid within 20 + 20 = 40 days of the beginning date of the discount period.

a. 3% b. 40 days

41. $2\frac{1}{2}$/10 - 20 Extra means that a $2\frac{1}{2}$% discount can be earned if the invoice is paid within 10 + 20 = 30 days of the beginning date of the discount period.

a. $2\frac{1}{2}$% b. 30 days

Exercise Set 6-4 Trade Discounts

1. Discount = \$100.00 • 0.20 — *Discount = List price • discount rate*

= \$20.00

5. Discount = \$510.50 • 0.16 — *Discount = List price • Discount rate*

= \$81.68

9. Total cost = \$12.50 • 10 — *Total cost = Cost per unit • Number of units*

= \$125.00

Discount = \$125.00 • 0.10 — *Discount = Total cost • Discount rate*

= \$12.50

13. Total cost = 12 · \$25.25 = \$303.00

Discount = \$303.00 · 0.14 = \$42.42

17. a. Discount = \$100.00 · 0.16 = \$16.00

b. Net cost = \$100.00 - 16.00 = \$84.00

21. a. Discount = \$6.12 · $0.33\frac{1}{3}$ = \$6.12($\frac{1}{3}$) = \$2.04

$33\frac{1}{3}\% = \frac{100}{300} = \frac{1}{3}$

b. Net cost = \$6.12 - 2.04 = \$4.08

25. a. Total cost = \$8.50 · 10 = \$85.00

b. Discount = \$85.00 · 0.20 = \$17.00

c. Net cost = \$85.00 - 17.00 = \$68.00

29. a. Total cost = \$46.20 · 22 = \$1016.40

b. Discount = \$1016.40 · 0.145 = \$147.38

c. Net cost = \$1016.40 - 147.38 = \$869.02

33. Discount = \$100.00 - 90.00   *Discount = List price - Net cost*

= \$10.00

Rate of discount = $\frac{\$10.00}{\$100.00}$ *Rate of discount = $\frac{\text{Discount}}{\text{List price}}$*

= 0.10 = 10%

37. Discount = \$238.00 - 193.97   *Discount = List price - Net cost*

= \$44.03

Rate of discount = $\frac{\$44.03}{\$238.00}$ *Rate of discount = $\frac{\text{Discount}}{\text{List price}}$*

= 0.185 = 18.5%

41. Net cost factor = (1.00 - 0.20)(1.00 - 0.10)

= (0.80)(0.90)

= 0.72

a. The net cost = \$120.00 · 0.72

= \$86.40

b. Discount = \$120.00 - 86.40

= \$33.60

45. Net cost factor $= (1.00 - 0.15)(1.00 - 0.10)$

$= (0.85)(0.90)$

$= 0.765$

a. Net cost $= \$265.00 \cdot 0.765$

$= \$202.72$ *Round down the net cost.*

b. Discount $= \$265.00 - 202.73$

$= \$62.28$ *Round up the discount.*

PART C Chapter Six Sample Test

In 1-5, state (a) the rate of discount, and (b) the number of days in the time period during which the discount can be earned.

			Answer
1.	2/20	1. a.	2%
		b.	20 days
2.	5/10 - 10 EX	2. a.	5%
		b.	20 days
3.	$3\frac{1}{2}$/15 - 5 X	3. a.	3½
		b.	20 days
4.	2/30 - 10 EX	4. a.	2%
		b.	40 days
5.	$4\frac{1}{2}$/10 ROG	5. a.	4½%
		b.	10 days after Rog

In 6-10, determine the final date the discount can be taken.

	Invoice Date	*Terms*	*Date Goods Received*		*Final Date of Discount*
6.	October 20	3/10 EOM	October 22	6.	11/10
7.	October 20	3/10 ROG	October 22	7.	11/1
8.	October 20	3/10-10X EOM	October 22	8.	11/20
9.	October 20	3/10-15 EX ROG	October 22	9.	11/16
10.	October 20	3/15-5X ROG	October 22 (31, 9)	10.	11/11

In 11-13, calculate (a) the amount of discount available, and (b) the amount due based on the date on which the invoice was paid.

	Invoice Amount	*Invoice Date*	*Terms*	*Date Goods Received*	*Date Invoice Paid*		
11.	$675.00	April 4	3/15 EOM	April 15	May 6	11. a.	20.25
						b.	654.25
12.	$7050	Sept 1	4/10 ROG	Sept 18	Sept 27	12. a.	282.00
						b.	6768.00
13.	$135,500	Dec 5	2/15-20X ROG	Dec 12	Jan 10	13. a.	2710.00
						b.	132790.00

In 14-16, calculate the net cost on each item with the given series discount.

	List Price	*Series Discount*	
14.	$250	15/10	14. ~~225.77~~ 191.25
15.	$8,410	15/10/5	15. 6111.97
16.	$15,600	20/15/5	16. 10077.60

In 17-20, solve each applied problem.

17. An invoice on furniture to Standard Furniture Company was dated April 10. The amount on the invoice was $72,800, and the terms were 3/10 N/30 EOM.

    a. Calculate the amount due if the invoice was paid May 8. 17. a. 70616

    b. Calculate the amount due if the invoice was paid May 20. b. 72800

18. Sampson's Auto Supply bought 50 cases of Premium Motor Oil at $16.80 per case. A volume discount of 15% is given on case lots.

    a. Compute the cost of the oil without a discount. 18. a. 840

    b. Compute the discount. b. 126

    c. Compute the net cost of the oil to Sampson's. c. 714

19. Fennel Hardware bought 20 Apex lawn mowers during the off-season. As a consequence they were given a 20/15 series discount. The list price on each mower was $120.

    a. Calculate the discount on the 20 mowers to Fennel Hardware. 19. a. ~~2400~~ 768

    b. Calculate the net cost of the mowers to Fennel Hardware. b. ~~1920~~ 1632

20. Hendricks Electrical Construction buys circuit breaker panels that cost $30 each from a distributor that offers an annual volume discount with the following schedule.

0-100	*0% discount*
101-300	*2% discount*
over 301	*5% discount*

As of December 28th Hendricks had purchased 275 of the panels, and consequently ordered 30 more to earn the higher discount.

    a. Calculate the discount on the order placed December 28th. 20. a. 45

    b. Calculate the discount on the previous panels purchased during the year. b. 165

CHAPTER SEVEN

THE MATHEMATICS OF SELLING

PART A Summary of Topics

OBJECTIVES

1. *Learn the definition of a markup.*
2. *Compute a markup based on cost.*
3. *Find the total cost of an item.*
4. *Find the markup of an item when total cost and selling price are known.*
5. *Compute the markup percent when markup based on cost and total cost are known.*
6. *Compute a markup based on selling price.*
7. *Compute the markup percent when markup based on selling price and selling price are known.*
8. *Find the selling price when total cost and markup percent are known.*
9. *Find the total cost when markup and markup percent are known.*
10. *Learn the definition of a markdown.*
11. *Compute a markdown.*
12. *Compute a new price.*
13. *Find the markdown percent.*
14. *Learn a description of sales tax.*
15. *Compute a sales tax for a taxable sale.*
16. *Find a taxable sale amount when the sales tax and tax rate are known.*
17. *Find the total amount of a sale, including the sales tax.*

SECTION 7-1 Markup Based on Cost

1. Markup is the dollar amount added to the cost of an item to determine a selling price.

   The Markup in the cost of an item is used to determine a selling price that is needed to pay for costs associated with the selling of the item, and the profit the merchant wants to make on the sale.

2. SP stands for Selling Price.

   C stands for Cost.

   M stands for Markup.

   $SP = C + M$ *Selling price = Cost + Markup.*

3. The following equations are alternate forms of this equation.

   $C = SP - M$ *Cost = Selling price - Markup*

   $M = SP - C$ *Markup = Selling price - Cost*

4. The markup is sometimes computed as a percent of the cost of the item. The symbol MC% will stand for the percent markup based on cost.

   $MC\% = \frac{M}{C}$ *Percent markup based on cost = $\frac{\text{Markup}}{\text{Cost}}$*

   Also, $M = C \cdot MC\%$ *Markup = Cost • Markup percent*

SECTION 7-2 Markup Based on Selling Price

1. A markup can also be determined as a percent of the selling price of the item.

   MS% stands for the markup percent based on selling price

   M stands for markup

   SP stands for selling price

   $MS\% = \frac{M}{SP}$ *Markup percent based on selling price* $= \frac{\textit{Markup}}{\textit{Selling price}}$

   Keep in mind that MC% stands for Markup percent based on Cost, and MS% stands for Markup percent based on Selling price. In general, for a given item, the two percents are different.

2. When the cost of an item is known, and the markup is a percent of the selling price, then the selling price can be calculated with the following equation.

   $$SP = \frac{C}{1 - MS\%}$$

   With this equation it is possible to find an unknown selling price by using the cost of the item, and the markup percent based on the selling price of the item.

3. When the selling price and the markup percent based on the selling price are known, then the cost can be determined with the following equation.

   $$C = SP - MS\% \cdot SP$$

   Since $MS\% = \frac{M}{SP}$ *and* $C = SP - M$, *then it follows that* $M = SP \cdot MS\%$, *and this expression can replace M in the equation* $C = SP - M$. *The result is the stated equation* $C = SP - MS\% \cdot SP$.

SECTION 7-3 Markdown

1. Markdown is the dollar amount subtracted from the selling price of an item to determine a new price.

   The Markdown in the selling price of an item is used for several reasons. Three of the more obvious reasons are:

   1. To increase sales.

   2. To reflect a decrease in cost from the manufacturer.

   3. To reduce inventory.

2. MD stands for Markdown.

   RP stands for Regular price (the Selling price before markdown).

   NP stands for New price (the reduced price of an item, sometimes called "sale price").

   MD% stands for Percent of markdown based on the regular price.

   $MD = RP - NP$ *Markdown = Regular price - New price*

3. The following equations are alternate forms of this equation.

NP = RP - MD *New price = Regular price - Markdown*

RP = NP + MD *Regular price = New price + Markdown*

4. If the Markdown is a percent of the Regular price, then

MD = RP • MD% *Markdown = Regular price • Markdown percent*

An alternate form of this equation is the following.

$MD\% = \frac{MD}{RP}$ *Markdown percent =* $\frac{\textit{Markdown}}{\textit{Regular price}}$

SECTION 7-4 Sales Tax

1. A sales tax is a tax levied on an item sold.

   A sales tax is one that is used by most states as a means of getting revenue. The tax rate can vary from state to state, and sometimes is different in cities within a given state because of a city sales tax.

   Not all items sold at the retail level are subject to a sales tax, thus the terms taxable sale and nontaxable sale.

2. TS stands for Taxable sale.

   ST stands for Sales tax.

   r stands for rate of sales tax

   ST = r • TS *Sales tax = rate of sales tax • Taxable sale*

3. The following equations are alternate forms of this equation.

   $r = \frac{ST}{TS}$ *rate of sales tax =* $\frac{\textit{Sales tax}}{\textit{Taxable sale}}$

   $TS = \frac{ST}{r}$ *Taxable sale =* $\frac{\textit{Sales tax}}{\textit{rate of sales tax}}$

4. AC stands for the Amount collected on a Taxable sale.

   AC = TS + ST *Amount collected = Taxable sale + Sales tax*

   The amount collected is the total cost of a taxable item that includes the price of the item plus the sales tax.

PART B Selected Solutions

Exercise Set 7-1 Markup Based on Cost

1. With C = $12.50 and M = $6.25

   SP = $12.50 + 6.25 *SP = C + M*

   = $18.75 *Selling price is $18.75.*

5. With $C = \$236.55$ and $M = \$106.45$

$SP = \$236.55 + 106.45$ — *$SP = C + M$*

$= \$343.00$ — *Selling price is \$343.00.*

9. a. With $C = \$85.50$ and $MC\% = 22\%$

$M = \$85.50 \cdot 0.22$ — *$M = C \cdot MC\%$*

$= \$18.81$ — *Markup is \$18.81.*

b. With $C = \$85.50$ and $M = \$18.81$

$SP = \$85.50 + 18.81$ — *$SP = C + M$*

$= \$104.31$ — *Selling price is \$104.31.*

13. With $SP = \$8.40$ and $M = \$2.10$

$C = \$8.40 - 2.10$ — *$C = SP - M$*

$= \$6.30$ — *Cost is \$6.30*

17. With $SP = \$384.69$ and $M = \$126.85$

$C = \$384.69 - 126.85$ — *$C = SP - M$*

$= \$257.84$ — *Cost is \$257.84.*

21. With $SP = \$189.98$ and $C = \$95.00$

$M = \$189.98 - 95.00$ — *$M = SP - C$*

$= \$94.98$ — *Markup is \$94.98.*

25. a. With $SP = \$75.00$ and $C = \$52.50$

$M = \$75.00 - 52.50 = \$22.50$ — *Markup is \$22.50.*

b. With $C = \$52.50$ and $M = \$22.50$

$MC\% = \frac{22.50}{52.50}$ — *$MC\% = \frac{M}{C}$*

$= 0.429$ — *To three decimal places.*

$= 42.9\%$ — *Written as a percent.*

29. a. With $SP = \$210.00$ and $M = \$42.00$

$C = \$210.00 - \$42.00 = \$168.00$ — *$C = SP - M$*

b. With $C = \$168.00$ and $M = \$42.00$

$MC\% = \frac{42.00}{168.00}$ — *$MC\% = \frac{M}{C}$*

$= 0.250$ — *To three decimal places.*

$= 25\%$ — *Written as a percent.*

Extended Calculator Exercises

1.

ENTER		PRESS	DISPLAY
Place 1 in the N register	1	N	1.00
The selling price	15.98	FV -	15.98
The cost	4.57	= PV	11.41
Calculate the markup percent based on cost		CPT %i	0.40

Exercise Set 7-2 Markup Based on Selling Price

1. With SP = \$25.00 and MS% = 20%

M = \$25.00 • 0.20 — *M = SP • MS%*

= \$5.00 — *Markup is \$5.00.*

5. With SP = \$520.00 and MS% = $33\frac{1}{3}$ %

M = \$520.00 • $\frac{1}{3}$ — *$33\frac{1}{3}\% = \frac{100}{300} = \frac{1}{3}$*

= \$173.33 — *Rounded to the nearest cent.*

9. With SP = \$90.00 and M = \$30.00

MS% = $\frac{30.00}{90.00}$ — *$MS\% = \frac{M}{SP}$*

= 0.333 — *To three decimal places*

= 33.3% — *To the nearest tenth of a percent*

As an exact percent, the answer to Exercise 9 can be written

$33\frac{1}{3}\%$.

13. With $C = \$6.75$ and $MS\% = 42\% = 0.42$

$$SP = \frac{\$6.75}{1 - 0.42} \qquad SP = \frac{C}{1 - MS\%}$$

$$= \frac{\$6.75}{0.58}$$

$= \$11.64$ — *Rounded to the nearest cent.*

17. With $C = \$72.50$ and $MS\% = 50\% = 0.50$

$$SP = \frac{\$72.50}{1 - 0.50} \qquad SP = \frac{C}{1 - MS\%}$$

$$= \frac{\$72.50}{0.50}$$

$= \$145.00$ — *Selling price is \$145.00.*

21. a. With $SP = \$215.00$ and $MS\% = 25\%$

$M = \$215.00 \cdot 0.25$ — $M = SP \cdot MS\%$

$= \$53.75$ — *Markup is \$53.75.*

b. With $SP = \$215.00$ and $M = \$53.75$

$C = \$215.00 - 53.75$ — $C = SP - M$

$= \$161.25$ — *Cost is \$161.25.*

25. With $SP = \$85.00$ and $MS\% = 24\%$

$M = \$85.00 \cdot 0.24$ — $M = SP \cdot MS\%$

$= \$20.40$ — *Markup is \$20.40.*

With $SP = \$85.00$ and $M = \$20.40$

$C = \$85.00 - 20.40$ — $C = SP - M$

$= \$64.60$ — *Cost is \$64.60.*

29. With $C = \$128$ and $MS\% = 36\% = 0.36$

$$SP = \frac{\$128}{1 - 0.36} \qquad SP = \frac{C}{1 - MS\%}$$

$$= \frac{\$128}{0.64}$$

$= \$200.00$ — *Selling price is \$200.00.*

With $SP = \$200.00$ and $C = \$128.00$

$M = \$200.00 - 128.00$ — $M = SP - C$

$= \$72.00$ — *Markup is \$72.00.*

33. With SP = \$2.79 and MS% = $33\frac{1}{3}$%

$M = \$2.79 \cdot \frac{1}{3}$ — *M = SP • MS%*

$= \$0.93$ — *Markup is \$0.93*

With SP = \$2.79 and M = \$0.93

$C = \$2.79 - 0.93$ — *C = SP - M*

$= \$1.86$ — *Cost is \$1.86.*

Exercise Set 7-3 Markdown

1. With RP = \$10.99 and NP = \$8.99

$M = \$10.99 - 8.99$ — *M = RP - NP*

$= \$2.00$ — *Markdown is \$2.00.*

5. With RP = \$9.99 and NP = \$8.99

$M = \$9.99 - 8.99$ — *M = RP - NP*

$= \$1.00$ — *Markdown is \$1.00.*

9. With RP = \$32.89 and MD% = 25%

$M = \$32.89 \cdot 0.25$ — *M = RP • MD%*

$= \$8.22$ — *Rounded to the nearest cent.*

13. With RP = \$9.98 and M = \$3.50

$NP = \$9.98 - 3.50$ — *NP = RP - M*

$= \$6.48$ — *New price is \$6.48.*

17. With RP = \$269.25 and M = \$45.26

$NP = \$269.25 - 45.26$ — *NP = RP - M*

$= \$223.99$ — *New price is \$223.99*

21. With RP = \$73.98 and MD% = $33\frac{1}{3}$%

a. $M = \$73.98 \cdot \frac{1}{3}$ — *M = RP • MD%*

$= \$24.66$ — *Markdown is \$24.66.*

b. $NP = \$73.98 - 24.66$ — *NP = RP - M*

$= \$49.32$ — *New price is \$49.32.*

25. a. With NP = \$16.97 and M = \$18.03

RP = \$16.97 + 18.03 — *RP = NP + M*

= \$35.00 — *Regular price is \$35.00.*

b. With RP = \$35.00 and M = \$18.03

$MD\% = \frac{\$18.03}{\$35.00}$ — $MD\% = \frac{M}{RP}$

= 0.515 — *Rounded to three decimal places.*

= 51.5% — *To the nearest tenth of a percent.*

29. a. With NP = \$90.00 and M = \$135.00

RP = \$90.00 + 135.00 — *RP = NP + M*

= \$225.00 — *Regular price is \$225.00*

b. With RP = \$225.00 and M = \$135.00

$MD\% = \frac{\$135.00}{\$225.00}$ — $MD\% = \frac{M}{RP}$

= 0.60 = 60% — *Markdown percent is 60%.*

35. With RP = \$2,325.98 and MD% = 5%

a. M = \$2,325.98 • 0.05 — *M = RP • MD%*

= \$116.30 — *Markdown is \$116.30.*

b. NP = \$2,325.98 - 116.30 — *NP = RP - M*

= \$2,209.68 — *New price is \$2,209.68.*

Exercise Set 7-4 Sales Tax

1. With TS = \$13.50 and r = 3%

ST = \$13.50 • 0.03 — *ST = TS • r*

= \$0.41 — *Rounded to the nearest cent.*

5. With TS = \$82.75 and r = 4.5%

ST = \$82.75 • 0.045 — *ST = TS • r*

= \$3.72 — *Rounded to the nearest cent.*

9. With TS = \$9.99 and r = 0.055

ST = \$9.99 • 0.055 — *ST = TS • r*

= \$0.55 — *Sales tax is \$0.55.*

13. Since 12.43 is between 12.38 and 12.62 in Appendix Table B-11, the sales tax is 0.50; that is, $0.50.

17. Since $226.75 = $200.00 + 26.75

Sales tax = $ 8.00 + 1.07

= $ 9.07

Notice, $200.00 • 0.04 = $8.00 and, 26.75 is between 26.63 and 26.87.

21. a. With TS = $18.75 and r = 3%

ST = $18.75 • 0.03 — *ST = TS • r*

= $0.56 — *Sales tax is $0.56.*

b. With TS = $18.75 and ST = $0.56

AC = $18.75 + 0.56 — *AC = TS + ST*

= $19.31 — *Amount collected is $19.31.*

25. a. With TS = $1,075.20 and r = 6%

ST = $1,075.20 • 0.06 — *ST = TS • r*

= $64.51 — *Sales tax is $64.51.*

b. With TS = $1,075.20 and ST = $64.51

AC = $1,075.20 + 64.51 — *AC = TS + ST*

= $1,139.71 — *Amount collected is $1,139.71.*

29. a. TS = $479.25 + 359.50 + 315.69

= $1,154.44 — *Taxable sale is $1,154.44.*

b. ST = $1,154.44 • 0.045

= $51.95 — *Sales tax is $51.95.*

c. AC = $1,154.44 + 51.95

= $1,206.39 — *Amount collected is $1,206.39.*

33. With Taxable sales of $308,144.03 and a tax rate of 5.25%,

Sales tax paid = $308,144.03 • 0.0525

= $16,177.56 — *Rounded to the nearest cent.*

PART C Chapter Seven Sample Test

In 1-3, find a. the markup, and b. the markup percent based on cost.

	Selling Price	Cost		
1.	$75.00	$60.00	1. a.	15.00
			b.	50%
2.	$218.10	$145.40	2. a.	72.70
			b.	50%
3.	$2824.00	$2,118.00	3. a.	706
			b.	33 1/3%

In 4-6, find a. the markup, and b. the selling price.

	Cost	MC%		
4.	$37.25	20%	4. a.	7.45
			b.	44.70
5.	$148.00	25.5%	5. a.	37.74
			b.	185.74
6.	$3,240	$33\frac{1}{3}$ %	6. a.	1080
			b.	4320

In 7-9, find a. the markup, and b. the markup percent based on the selling price to the nearest tenth of a percent.

	Selling Price	Cost		
7.	$7.95	$6.36	7. a.	1.59
			b.	20%
8.	$132.50	$102.82	8. a.	29.68
			b.	22.4%
9.	$4,650	$4,262.50	9. a.	378.50
			b.	8 1/3%

In 10-12, find the selling price.

	Cost	MS%		
10.	$8.40	30%	10.	12.00
11.	$217.50	$33\frac{1}{3}$ %	11.	326.25
12.	$3,540.40	16.5%	12.	4240.00

In 13-15, find a. the markdown, and b. the new price.

	Regular Price	*MD%*		
13.	$22.50	10%	13. a.	2.25
			b.	20.25
14.	$375.00	24%	14. a.	93.75
			b.	281.25
15.	$14,996.00	7.5%	15. a.	1124.70
			b.	13871.30

In 16-18, use the given information to fill in the blanks. Round dollar amounts to the nearest cent, and percents to the nearest tenth of a percent.

	Cost	*Markup*	*Selling Price*	*MC%*	*MS%*
16.	$12.00	6.00	$18.00	50%	33 1/3%
17.	20.00	$5.00	$25.00	25%	20%
18.	182.00	28.00	$210.00	15.4%	$13\frac{1}{3}$ %

cost % ↑ & cost = cost ↑ = 3.52

19. To determine a selling price for paint, Smith Hardware uses a 40% markup based on cost. If the markup on a gallon of house paint is $3.52, find

    a. the cost — 19. a. 8.80

    b. the selling price — b. 12.32

C + MC = ₵C + MC% SP =

20. What is the cost of a boy's jacket which sells for $39.50, if the markup based on selling price is 32%? — 20. 26.86

21. The regular price of a woman's necklace is $148.00. If the new price is $111.00, compute the markdown percent. — 21. 25%

22. A horse blanket is on sale for $59.50. If the blanket was discounted $17.95, find the regular selling price. — 22. 77.45

In 23-25, compute a. the sales tax, and b. the amount collected.

	Taxable Sale	*Sales Tax Rate*		
23.	$21.80	4%	23. a.	.87
			b.	22.67
24.	$132.60	5.5%	24. a.	7.29
			b.	139.89
25.	$5,620	6%	25. a.	337.20
			b.	5957.20

CHAPTER EIGHT

SIMPLE INTEREST

PART A Summary of Topics

OBJECTIVES

1. *Learn to compute simple interest.*
2. *Compute simple interest using the exact time and exact interest method.*
3. *Compute simple interest using the approximate time and ordinary interest method.*
4. *Find simple interest using the 6%-60 day method.*
5. *Find the rate when the interest, principal, and time are known.*
6. *Find the time when the interest, principal, and rate are known.*
7. *Find the principal when the interest, rate, and time are known.*
8. *Learn what is a promissory note.*
9. *Learn what is a noninterest-bearing note.*

SECTION 8-1 Simple Interest

1. Interest is a fee that is paid for the use of money.

   Two methods are commonly used to compute the interest when money is borrowed:

   a. The simple interest method, which is studied in Chapter 8.

   b. The compound interest method, which is studied in Chapter 9.

2. When the simple interest method is used to compute the fee for borrowing money, there are three factors that determine the amount of the fee:

   Factor 1. The Principal, which is the amount of money borrowed.

   Factor 2. The Rate, which is the percent used to compute the interest.

   Factor 3. The Time, which is the number of years for which the money is borrowed.

3. If I is the interest, P is the principal, R is the rate, and T is the time in years, then

   $$I = P \cdot R \cdot T$$ *Interest = (Principal)(Rate)(Time)*

4. Simple interest can be Exact or Ordinary.

   Exact interest uses a 365 day calendar year, or 366 days on leap years.

   Ordinary interest uses a 360 day year.

5. Time can be calculated as Exact or Approximate.

   Exact time counts every day, except the first, for which the loan is made.

   Approximate time assumes that each month (including February) has 30 days.

SECTION 8-2 The 6%-60 Day Rule

1. If x dollars is borrowed at 6% annual interest rate for 60 days, then the ordinary interest can be determined by moving the decimal point in x two places to the left.

   The 6%-60 day rule applies whether the 60 days is exact time or approximate time. However, the interest must be computed using the ordinary interest method so that the denominator of the time fraction is 360, and not 365.

2. Steps 1 and 2 can be used to adjust the 6%-60 day rule when the rate is 6% and the time is y days:

   Step 1. Move the decimal point in x two places to the left.

   Step 2. Multiply the value in Step 1 by $\frac{y}{60}$.

3. Steps 1 and 3 can be used to adjust the 6%-60 day rule when the rate is z% and the time is 60 days.

   Step 1. Move the decimal point in x two places to the left.

   Step 3. Multiply the value in Step 1 by $\frac{z}{6}$.

4. Steps 1, 2 and 3 can be used to adjust the 6%-60 day rule when the time is y days and the rate is z%.

   Step 1. Move the decimal point in x two places to the left.

   Step 2. Multiply the value in Step 1 by $\frac{y}{60}$.

   Step 3. Multiply the product in Step 2 by $\frac{z}{6}$.

SECTION 8-3 Finding the Principal, Rate, and Time

The equation that defines the relationship between simple interest, principal, rate and time is

Interest = Principal · Rate · Time $\qquad I = P \cdot R \cdot T$

1. If the rate is the only unknown quantity in the simple interest equation, then

   $$\text{Rate} = \frac{\text{Interest}}{\text{Principal} \cdot \text{Time}} \qquad R = \frac{I}{P \cdot T}$$

2. If the time is the only unknown quantity in the simple interest equation, then

   $$\text{Time} = \frac{\text{Interest}}{\text{Principal} \cdot \text{Rate}} \qquad T = \frac{I}{P \cdot R}$$

3. If the principal is the only unknown quantity in the simple interest equation, then

   $$\text{Principal} = \frac{\text{Interest}}{\text{Rate} \cdot \text{Time}} \qquad P = \frac{I}{R \cdot T}$$

SECTION 8-4 Promissory Notes

1. A promissory note is a promise in writing made by one person to another, agreeing to pay on demand (or at a definite time) a specified amount of money.

If the time on a note is less than one year, then exact time is used to determine the due date. However, in the equation $I = P \cdot R \cdot T$, the denominator of T is 360. Thus the ordinary interest method is used to calculate the interest.

2. If the interest on a promissory note is deducted from the principal on the note before the money is given to the borrower, the term noninterest-bearing note can be used. The amount the borrower gets is called the proceeds.

$$\text{Proceeds} = \text{Principal} - \text{Interest}$$

The term noninterest-bearing note may be a bit deceptive, since the note does include a finance charge to borrow the money. However, only the principal is paid on the due date, since the interest was deducted at the time the note was initiated.

PART B Selected Solutions

Exercise Set 8-1 Simple Interest

1. With P = \$7,000, R = 12% and T = 2 years

$I = \$7,000 \cdot 0.12 \cdot 2$ — $I = P \cdot R \cdot T$

$= \$1,680$ — *The interest is \$1,680.*

5. With P = \$4,000, R = 8.75% and T = 2 years

$I = \$4,000 \cdot 0.0875 \cdot 2$ — $I = P \cdot R \cdot T$

$= \$700$ — *The interest is \$700.*

9. a. With P = \$1,400, R = 14% and $T = \frac{90}{365}$

$I = \$1,400 \cdot 0.14 \cdot \frac{90}{365}$ — $I = P \cdot R \cdot T$

$= \$48.33$ — *Exact interest is \$48.33.*

b. With P = \$1,400, R = 14% and $T = \frac{90}{360}$

$I = \$1,400 \cdot 0.14 \cdot \frac{90}{360}$ — $I = P \cdot R \cdot T$

$= \$49.00$ — *Ordinary interest is \$49.00.*

13. a. With P = \$750, R = 12.5% and $T = \frac{360}{365}$

$I = \$750 \cdot 0.125 \cdot \frac{360}{365}$ — $I = P \cdot R \cdot T$

$= \$92.47$ — *Exact interest is \$92.47*

b. With P = \$750, R = 12.5% and $T = \frac{360}{360}$

$I = \$750 \cdot 0.125 \cdot 1$ — $I = P \cdot R \cdot T$

$= \$93.75$ — *Ordinary interest is \$93.75*

17. a. For exact time,

August 12 is day 224		*Using the values in Table 8-1.*
June 24 is day 175		
The total is 49 days		

b. For approximate time,

June 24 - June 30 -	6 days	*Every month in the year has 30 days.*
July 01 - July 30 -	30 days	
August 01 - August 12 -	12 days	
The total is	48 days	

21. a. For exact time and a leap year,

December 3 is day 338		*Using the values in Table 8-1.*
January 3 is day 3		
The total is 335 days		

b. For approximate time,

January 3 - January 30	27 days	*Every month in the year has 30 days.*
February 1 - November 30 (10 months)	300 days	
December 1 - December 3	3 days	
The total is	330 days	

25. a. For exact time and a leap year,

December 5 - December 31	26 days	
January 1 - January 31	31 days	*February has 29 days since it is a leap year.*
February 1 - February 29	29 days	
March 1 - March 9	9 days	
The total is	95 days	

b. For approximate time,

December 5 - December 30	25 days
January 1 - January 30	30 days
February 1 - February 30	30 days
March 1 - March 9	9 days
The total is	94 days

29. a. From Table 8-1,

February 19 is day 50	
January 9 is day 9	
The total is 41 days	*Exact time*

With P = \$1,500, R = 12% and $T = \frac{41}{365}$

$I = \$1,500 \cdot 0.12 \cdot \frac{41}{365}$

= \$20.22, to the nearest cent *Exact interest, using exact time.*

b. From January 9 - January 30 21 days

February 1 - February 19 19 days

The total is 40 days *Approximate time*

With P = \$1,500, R = 12% and $T = \frac{40}{360}$

$I = \$1,500 \cdot 0.12 \cdot \frac{40}{360}$

= \$20.00 *Ordinary interest, using approximate time*

33. a. The exact time from August 13 to October 7 is 55 days.

With P = \$1,640, R = $10\frac{3}{4}$ % and $T = \frac{55}{365}$

$I = \$1,640 \cdot 0.1075 \cdot \frac{55}{365}$

= \$26.57 *Exact interest, using exact time.*

b. The approximate time from August 13 to October 7 is 54 days.

$I = \$1,640 \cdot 0.1075 \cdot \frac{54}{360}$

= \$26.45 *Ordinary interest, using approximate time.*

37. With P = \$3,200, R = 7.5% and T = 2.5 years

I = \$3,200 · 0.075 · 2.5

= \$600.00 *Exact interest, using exact time.*

<u>Exercise Set 8-2</u> <u>The 6%-60 Day Rule</u>

1. With P = \$580, R = 6% and T = 60 days,

I = \$5.80 *Move the decimal point in P two places to the left.*

5. With P = \$586.40, R = 6% and T = 60 days,

I = \$5.86, to the nearest cent.

9. With P = \$1,200, R = 6% and T = 20 days,

<u>Step 1.</u> I = \$12.00 *The interest for 60 days.*

<u>Step 2.</u> $I = \$12.00 \cdot \frac{20}{60}$ *The time is y = 20 days.*

= \$4.00 *The interest for 20 days.*

13. With P = \$7,140.50, R = 6% and T = 135 days,

Step 1. I = \$71.405 — *The interest for 60 days.*

Step 2. $I = \$71.405 \cdot \frac{135}{60}$ — *The time is y = 135 days.*

= \$160.66 — *The interest for 135 days.*

17. With P = \$650, R = 12% and T = 60 days,

Step 1. I = \$6.50 — *The interest at 6%.*

Step 3. $I = \$6.50 \cdot \frac{12}{6}$ — *The rate is z% = 12%.*

= \$13.00 — *The interest at 12%.*

21. With P = \$4,510, R = 8.5% and T = 60 days,

Step 1. I = \$45.10 — *The interest at 6%.*

Step 3. $I = \$45.10 \cdot \frac{8.5}{6}$ — *The rate is z% = 8.5%.*

= \$63.89 — *The interest at 8.5%.*

25. With P = \$632, R = 9.0% and T = 100 days,

Step 1. I = \$6.32 — *The interest at 6% for 60 days.*

Step 2. $I = \$6.32 \cdot \frac{100}{60}$ — *The interest adjusted for y = 100 days.*

Step 3. $I = \$6.32 \cdot \frac{100}{60} \cdot \frac{9.0}{6}$ — *The interest adjusted for z% = 9%.*

= \$15.80 — *The ordinary interest.*

Notice that the product in Step 2 is not rounded. In general, do not round intermediate values, but only the final answer.

29. With P = \$932.20, R = 15.0% and T = 15 days,

Step 1. I = \$9.322

Step 2. $I = \$9.322 \cdot \frac{15}{60}$

Step 3. $I = \$9.322 \cdot \frac{15}{60} \cdot \frac{15.0}{6}$

= \$5.83

Exercise Set 8-3 Finding the Principal, Rate, and Time

1. With P = \$1,000, T = 60 days and I = \$10.00,

$$R = \frac{\$10.00}{\$1,000 \cdot \frac{60}{360}}$$ — $R = \frac{I}{P \cdot T}$

= 0.06 — *Ordinary interest is used, thus the denominator is 360.*

= 6%

5. With P = \$2,700, T = 72 days,

and I = \$54,

$$R = \frac{\$54}{\$2,700 \cdot \frac{72}{360}} \qquad R = \frac{I}{P \cdot T}$$

= 0.10

= 10%

9. With P = \$1,280, R = 12%

and I = \$76.80,

$$T = \frac{\$76.80}{\$1,280 \cdot 0.12} \qquad T = \frac{I}{P \cdot R}$$

= 0.5 *$\frac{1}{2}$ of 360 days*

= 180 days

13. With P = \$720, R = 14.6%,

and I = \$35.04,

$$T = \frac{\$35.04}{\$720 \cdot 0.145} \qquad T = \frac{I}{P \cdot R}$$

$= \frac{1}{3}$ *$\frac{1}{3}$ of 360 days*

= 120 days

17. With R = 9%, T = 2 years,

and I = \$810,

$$P = \frac{\$810}{0.09 \cdot 2} \qquad P = \frac{I}{R \cdot T}$$

= \$4,500

21. With R = 14.5%, T = 72 days, and I = \$87,

$$P = \frac{\$87}{0.145 \cdot \frac{72}{360}} \qquad P = \frac{I}{R \cdot T}$$

= \$3,000 *Since ordinary interest is computed, the denominator is 360.*

Exercise Set 8-4 Promissory Notes

1. On the promissory note in Figure 8-A,

   a. the maker is Elk Meadows Resort, Inc. *The borrower*

   b. the payee is Big Bend City Bank *The lender*

   c. the principal is \$11,750

   d. the rate of interest is 10.5% *R = 10.5%*

   e. the term of the note is 100 days *T = 100 days*

5. a. With I = \$170.67, R = 8% and T = 100 days

$$P = \frac{\$170.67}{0.08 \cdot \frac{100}{360}} \qquad P = \frac{I}{R \cdot T}$$

= \$7,680.15 *To the nearest dollar, P = \$7,680.*

   b. From Table 8-1, March 10 is day 69. The term of the note is 100 days, and 69 + 100 = 169. In Table 8-1, June 18 is day 169. Thus the due date is June 18th.

9. With P = \$2,600, I = \$910, and R = 8.75%,

$$T = \frac{\$910}{\$2,600 \cdot 0.0875} \qquad T = \frac{I}{P \cdot R}$$

= 4 years

13. a. With P = \$320,000, R = 13.1%, and T = 75 days,

$$I = \$320{,}000 \cdot 0.131 \cdot \frac{75}{360} \qquad \textit{I = P · R · T}$$

$$= \$8{,}733.33 \qquad \textit{Rounded to the nearest cent}$$

b. $\text{Proceeds} = \$320{,}000 - 8{,}733.33$ *Proceeds = Principal - Interest*

$= \$311{,}266.67$,

PART C Chapter Eight Sample Test

In 1-5, for each pair of dates find (a) the exact number of days and (b) the approximate number of days.

1. April 10 to November 15 (same year) — 1. a. 219 b. 215

2. May 20 to July 10 (same year) — 2. a. 51 b. 51

3. January 15 to October 28 (leap year) — 3. a. 287 b. 284

4. November 5 to January 18 (following year) — 4. a. 74 b. 73

5. February 7 to August 6 (not a leap year) — 5. a. 180 b. 177

In 6-10, find the unknown interest, principal, rate or time. Use the ordinary interest method for computing the answers. Round dollar amounts to the nearest cent, percent answers to the nearest tenth of a percent, and time answers to a fractional part of a year.

6. The principal is $1500, the rate is 12.3%, and the time is $3\frac{1}{2}$ years. — 6. $645.75

7. The principal is $24,900, the rate is 10.8%, and the time is 120 days. — 7. $896.40

8. The interest is $3060, the rate is 8.5%, and the time is 5 years. — 8. $7200.00

9. The interest is $73, the principal is $750, and the time is 240 days. — 9. 14.6%

10. The interest is $3240, the principal is $9600, and the rate is 13.5%. — 10. 2½ years

In 11-15, use the 6%-60 Day Rule to compute the interest to the nearest cent.

	Principal	Rate	Time		
11.	$825.00	6%	60 days	11.	8.28
12.	$1375.00	6%	45 days	12.	10.31
13.	$6,910.00	10%	60 days	13.	115.17
14.	$650.00	13.5%	90 days	14.	21.94
15.	$4,980.00	12.8%	150 days	15.	265.60

16. Tina Washington issued a promissory note with a face value of $3,500, a stated interest rate of 11.5%, and a period of 219 days. Find the exact interest Tina must pay on the note.

16. 244.85

241.50

17. Barb and Ira David maintain a fund of money on which their children can borrow money for short periods of time. Their daughter borrowed $2,175.00 from the fund for 9 months and repaid the fund $2,344.65. What is the rate of interest on the fund?

3/4

1631.25

17. 10.4%

18. Gary Chan needs to borrow $4,000 for 540 days. He can get the money for an interest rate of 13.5%.

a. Compute the exact interest on the loan.

b. Compute the ordinary interest on the loan.

18. a. 798.90

b. 810.00

19. Vickie Fuller has $7,500 in a savings account that pays interest at a rate of 7.8%. How long will it take this account to yield $1462.50 interest?

19. 2.5 years

20. Ruberstein Bakery Shop needs to borrow $36,000 to remodel the kitchen part of the shop. They offer a non-interest bearing note at 11.8% interest and 1.5 years time.

a. Compute the interest on the note.

b. Compute the proceeds.

20. a. 6372

b. ~~56628~~

29628

CHAPTER NINE

COMPOUND INTEREST AND PRESENT VALUE

PART A Summary of Topics

OBJECTIVES

1. *Learn what is meant by compound interest.*
2. *Compute the compound interest on a principal amount by using the multistep method.*
3. *Compute compound interest by the formula method.*
4. *Compute compound interest by the table method.*
5. *Learn what is meant by present value.*
6. *Identify the symbols used in the formulas for present value.*
7. *Find the relationship between present value and compound interest.*
8. *Find the future value of a lump sum deposit using the formula or present-value tables method.*
9. *Study other applications of present value.*
10. *Learn what is meant by an amortized loan.*
11. *Compute the principal and interest portions of a payment on an amortized loan.*
12. *Compute the interest on an amortized loan when the principal and payments are known.*
13. *Compute the payment on an amortized loan.*
14. *Estimate the interest portion of a payment using the rule of 78.*
15. *Learn what is meant by an annuity.*
16. *Learn what is a single-premium annuity.*
17. *Learn what is an increasing annuity.*
18. *Compute the present value of an annuity.*
19. *Learn what is meant by a sinking fund.*

SECTION 9-1 Compound Interest

1. When interest is compounded, then interest for one time period is automatically added to the principal, and the interest for the next period is computed on the new principal (= old principal + interest).

   Conversion period (or period) - the time between the computations of interest. The rate of interest is r, to distinguish from annual interest rate R.

2. To compute the interest for one conversion period, the following equation can be used:

   $I = P \cdot r$ — P is the principal for that particular period.
   r is the interest rate for any conversion period.

   When interest is compounded, then the principal changes for succeeding periods. Notice, if the interest is removed at the end of each conversion period and not allowed to accumulate to the principal, then the interest would appear to be the same as simple interest. When interest from one interest period to the next is added to the principal (and subsequently earn interest also), then the phenomenon of compound interest takes place.

3. A stands for the compound amount

P stands for the original principal

r stands for the interest rate per period

n stands for the number of periods

$$A = P(1 + r)^n$$

The values for $(1 + r)^n$ can be found in the Present Value and Future Value Tables in Appendix A for many values of r and n. The table value, called Table Factor, will be found in Column I, headed "How $1 left at Compound Interest will grow".

SECTION 9-2 An Introduction to Present Value

1. In the formulas for present value, the following symbols are used:

P stands for the Principal amount	*This is the same P that is in the compound interest formula.*
PMT stands for Payment	*Payment occurs in formulas for amortized loans, and annuities.*
PV stands for the Present Value	*Usually refers to the present value of an amount of money to be received in the future.*
FV stands for the Future Value	*Usually refers to the future value of an amount of money in the present.*
r stands for a Periodic Interest Rate	*This is the same r that is in the compound interest formula.*
n stands for the Number of Conversion Periods	*This is the same n that is in the compound interest formula.*

2. Letting A be the future value of an amount of money P in the present, and P be the present value of an amount of money A to be received in the future, then the formula

$$A = P(1 + r)^n$$

becomes,

$$FV = PV(1 + r)^n$$

3. The future value of a lump sum deposit of an amount of money in a compound interest fund can be computed with the formula,

$$FV = PV(1 + r)^n$$

The Present Value Tables in the Appendix can be used for selected values of r. The Table Factor is found in Column I, similar to the manner used in Section 9-1.

4. When the formula for FV is solved for PV, the following formula is obtained.

$$PV = FV \frac{1}{(1 + r)^n}$$

This formula can be used to determine the Present Value of an amount of money in the future after n periods of compounding interest with periodic interest rate r. The Table Factors in the Present Value Tables are found in Column II for the expression $\frac{1}{(1 + r)^n}$.

SECTION 9-3 Amortized Loans

1. Amortization is the process of repaying a loan by making a series of equal periodic payments that include both principal and interest.

   When a loan is amortized, each payment includes a portion that applies to the Principal, and the other part pays the interest for the previous conversion period. This is the manner in which a home mortgage is set up. Each payment on the loan pays the interest since the previous payment, plus an amount that reduces the balance of the loan on the house.

2. The interest on an amortized loan can be calculated when the principal amount of the loan is known, and the amount and number of the periodic payments are known.

   Interest = amount in payments - the principal amount

3. If PV stands for the amount to be financed, r is the periodic interest rate, and n is the number of payments for the loan, then

$$PMT = PV\left(\frac{r}{1 - \frac{1}{(1 + r)^n}}\right)$$

   Use the following steps when working the formula with a calculator.

   Step 1. *Compute $(1 + r)^n$ and do not round the number.*

   Step 2. *Divide 1 by the value obtained in Step 1.*

   Step 3. *Subtract the number in Step 2 from 1.*

   Step 4. *Divide r by the difference obtained in Step 3.*

   Step 5. *Multiply the PV by the quotient obtained in Step 4.*

4. The Rule of 78 applies to amortized loans of one year. The fractional part of interest included in each of the twelve payments is in the reverse order of the number of the payment. That is,

   Payment 1 $\frac{12}{78}$ of the payment is interest

   Payment 2 $\frac{11}{78}$ of the payment is interest

   and so on, until

   Payment 12 $\frac{1}{78}$ of the payment is interest.

   For loans of two years, a Rule of 300 could be used. That is,

   Payment 1 $\frac{24}{300}$ *of the payment is interest*

   Payment 2 $\frac{23}{300}$ *of the payment is interest*

   and so on, until

   Payment 24 $\frac{1}{300}$ *of the payment is interest.*

SECTION 9-4 Annuities and Sinking Funds

1. An annuity is a fund or investment into which periodic deposits are made or from which periodic payments are withdrawn.

   A. *With an ordinary annuity, each deposit is made at the end of the periodic payment schedule.*

   B. *With an annuity due, each deposit is made at the beginning of the periodic payment schedule.*

2. When regular periodic payments are taken from an annuity, then the amount of each payment depends on the amount in the annuity (PV), the periodic interest rate at which the balance in the annuity is earning interest (r), and the number of payments to be taken from the fund (n). The payments can be determined with the same formula used to calculate the payments on an amortized loan.

$$PMT = PV\left(\frac{r}{1 - \frac{1}{(1+r)^n}}\right)$$

For selected values of r and n, a Table Factor can be found in Column III in the tables in Appendix A. For such values, the formula becomes:

$$PMT = PV \cdot \text{Table Factor}$$

3. An increasing annuity is one into which regular payments (PMT) of the same amount are deposited. An IRA (Individual Retirement Account) is an example of an increasing annuity.

4. The future value (FV) of an increasing annuity with regular payments (PMT) of the same amount, periodic interest rate r, and n periods, can be calculated with the following equation:

$$FV = PMT\left(\frac{(1+r)^n - 1}{r}\right)$$

For the selected values of r and n in the tables in Appendix A, the Table Factor for this equation is found in Column V. For such values of r and n found in these tables, the formula becomes:

$$FV = PMT \cdot \text{Table Factor}$$

5. The present value (PV) of a series of payments (PMT) to be received, with r the inflation rate per period, and n the number of periods, can be calculated with the following equation:

$$PV = PMT(1+r)\left(\frac{1 - \frac{1}{(1+r)^n}}{r}\right)$$

For the selected values of r and n in the tables in Appendix A, the Table Factor for this equation is found in Column IV. For such values of r and n found in these tables, the formula becomes:

$$PV = PMT(1+r) \cdot \text{Table Factor}$$

6. A sinking fund is an account into which periodic payments are made to accumulate a certain amount of money at the end of some specified period of time.

With the sinking fund concept, a lump sum is needed at some time in the future. Then periodic payments are deposited in a fund that earns compound interest. At the end of the specified time, the necessary lump sum is in the fund.

7. The periodic payments (PMT) needed to accumulate a sinking fund with a future value (FV), at a specified periodic interest rate (r) , and number of periods (n), can be calculated with the following equation:

$$PMT = FV\left(\frac{r}{(1+r)^n - 1}\right)$$

For the selected values of r and n in the tables in Appendix A, the Table Factor for this equation is found in Column VI. For such values of r and n found in these tables, the formula becomes:

$$PMT = FV \cdot \text{Table Factor}$$

PART B Selected Solutions

Exercise Set 9-1 Compound Interest

1. With R = 0.06 and four periods, $r = \frac{0.06}{4} = 0.015$.

First quarter: I = \$5,000(0.015) = \$75

Principal for Second quarter is \$5,075.

Second quarter: I = \$5,075(0.015) = \$76.125

Principal for Third quarter is \$5,151.125

Third quarter: I = \$5,151.125(0.015) = \$77.266875

Principal for Fourth quarter is \$5,228.391875

Fourth quarter: I = \$5,228.391875(0.015) = \$78.42587813

The amount at the end of the fourth quarter is \$5,306.817753

To the nearest cent, A = \$5,306.82

5. With R = 0.12 and twelve periods, $r = \frac{0.12}{12} = 0.01$

First month: I = \$15,000(0.01) = \$150

Second month: I = \$15,150(0.01) = \$151.50

Third month: I = \$15,301.50(0.01) = \$153.015

Fourth month: I = \$15,454.515(0.01) = \$154.54515

Fifth month: I = \$15,609.06015(0.01) = \$156.0906015

Sixth month: I = \$15,765.15075(0.01) = \$157.6515075

Seventh month: I = \$15,922.80226(0.01) = \$159.2280226

Eighth month: I = \$16,082.03028(0.01) = \$160.8203028

Ninth month: I = \$16,242.85058(0.01) = \$162.4285058

Tenth month: I = \$16,405.27909(0.01) = \$164.0527909

Eleventh month: I = \$16,569.33188(0.01) = \$165.6933188

Twelfth month: I = \$16,735.0252(0.01) = \$167.350252

Amount at end of twelfth month is \$16,902.37545, which is \$16,902.38 to the nearest cent.

9. For one year and semiannual periods, n = 2. With P = \$780.00,

$r = \frac{0.13}{2} = 0.065,$

$A = \$780.00(1 + 0.065)^2$ $\qquad A = P(1 + r)^n$

$= \$780.00(1.134225)$

$= \$884.70$, to the nearest cent

13. For three years and semiannual periods, n = 6. With P = \$100.00, and

$r = \frac{0.12}{2} = 0.06,$

a. $A = \$100.00(1 + 0.06)^6$ $\qquad A = P(1 + r)^n$

$= \$100.00(1.418519112)$

$= \$141.85$, to the nearest cent

b. $I = \$141.85 - 100.00$ $\qquad I = A - P$

$= \$41.85$

17. For two years and monthly periods, n = 24. With P = \$1,000, and

$r = \frac{0.06}{12} = 0.005,$

a. $A = \$1,000(1 + 0.005)^{24}$ $\qquad A = P(1 + r)^n$

$= \$1,000(1.127159776)$

$= \$1,127.16$, to the nearest cent

b. $I = \$1,127.16 - 1,000.00$ $\qquad I = A - P$

$= \$127.16$

21. For $2\frac{1}{4}$ years and quarterly periods, n = 9. With P = \$25,000,

and $r = \frac{0.10}{4} = 0.025,$

a. $A = \$25,000(1 + 0.025)^9$ $\qquad A = P(1 + r)^n$

$= \$25,000(1.24886297)$

$= \$31,221.57$, to the nearest cent

b. $I = \$31,221.57 - 25,000.00$ $\qquad I = A - P$

$= \$6,221.57$

25. With R = 8% and semiannual periods, $r = \frac{8\%}{2} = 4\%$.

With 10 years and semiannual periods, n = 20.

From Table A-26, Column I, the Table Factor is 2.19112315.

$A = \$4,000(2.19112315)$ $\qquad A = P \cdot \text{Table Factor}$

$= \$8764.49$, to the nearest cent

Notice, $(1.04)^{20} = 2.19112315$ *to eight decimal places.*

29. With R = 9% and monthly periods, $r = \frac{9\%}{12} = 0.75\%$.

With $2\frac{1}{2}$ years and monthly periods, n = 30.

From Table A-5, Column I, the Table Factor is 1.25127178

A = $9,450(1.25127178) *A = P • Table Factor*

= $11,824.52, to the nearest cent

Notice, $(1.0075)^{30}$ = 1.25127178 to eight decimal places.

33. a. With $r = \frac{6\%}{12} = 0.5\%$ and n = 12,

A = $10,000(1.06167781) *Appendix Table A-2*

= $10,616.8, to the nearest cent

I = $616.78

b. With $r = \frac{6.4\%}{2} = 3.2\%$ and n = 2,

A = $\$10{,}000(1.032)^2$ *$A = P(1 + r)^n$*

= $10,650.24, to the nearest cent

I = $650.24

c. Fund B yields $33.46 more interest than Fund A.

Exercise Set 9-2 An Introduction to Present Value

1. With PV = $500, $r = \frac{12\%}{2} = 6\%$, and n = 14,

FV = $\$500(1 + 0.06)^{14}$ *$FV = PV(1 + r)^n$*

= $500(2.26090398) *Table Factor in Column I of Table A-32.*

= $1,130.45, to the nearest cent

5. With PV = $19,500, $r = \frac{11\%}{12}$ = 0.9166667%, and n = 24,

FV = $\$19{,}500(1 + 0.9166667)^{24}$ *$FV = PV(1 + r)^n$*

= $19,500(1.24482854) *Table Factor in Column I of Table A-7.*

= $24,274.16, to the nearest cent

9. With PV = $24,200, $r = \frac{7.5\%}{4} = 1.875\%$, and n = 60,

FV = $\$24{,}200(1 + 0.01875)^{60}$ *$FV = PV(1 + r)^n$*

= $24,200(3.04829719) *To eight decimal places*

= $73,768.79, to the nearest cent

13. With FV = $1,600, $r = \frac{6\%}{12} = 0.5\%$, and n = 18,

PV = $\$1{,}600 \frac{1}{(1 + 0.05)^{18}}$ *$PV = FV \frac{1}{(1 + r)^n}$*

= $1,600(0.91413616) *Table Factor in Column II of Table A-2.*

= $1,462.62

17. With FV = \$26,850, $r = \frac{10\%}{2} = 5\%$, and n = 8,

$PV = \$26,850 \frac{1}{(1 + 0.05)^8}$ — *$PV = FV \frac{1}{(1 + r)^n}$*

$= \$26,850(0.67683936)$ — *Table Factor in Column II of Table A-30.*

= \$18,173.14, to the nearest cent

21. With FV = \$8,125, $r = \frac{14\%}{12} = 1.166667\%$, and n = 360,

PV = \$8,125 (Table Factor) — *PV = FV • Table Factor*

= \$8,125(0.01536458) — *Table Factor in Column II of Table A-10.*

= \$124.84

Notice that \$124.84 deposited today in a fund that has an annual interest rate of 14% and compounds interest monthly will grow to \$8,125 in 30 years.

25. a. With PV = \$75,000, $r = \frac{9.25\%}{4} = 2.3125\%$, and n = 20,

$FV = \$75,000 \frac{1}{(1 + 0.023125)^{20}}$

= \$75,000(1.57969752), to eight decimal places.

= \$118,477.31, to the nearest cent.

b. Interest = \$118,477.31 - \$75,000.00

= \$43,477.31

Exercise Set 9-3 Amortized Loans

1. Payment 1

With P = \$6,500, and $r = \frac{12\%}{12} = 1\%$

I = \$6,500 • 0.01 = \$65.00 — *I = P • r*

Amount applied to principal = \$577.52 - \$65.00 — *Payment - Interest*

= \$512.52

Payment 2

With P = \$6,500.00 - 512.52 = \$5,987.48, and r = 0.01,

I = \$5,987.48 • 0.01 = \$59.87 — *Interest*

Amount applied to principal = \$577.52 - \$59.87

= \$517.65 — *Principal*

Payment 3

With P = \$5,987.48 - 517.65 = \$5,469.83, and r = 0.01

I = \$5,469.83 • 0.01 = \$54.70 — *Interest*

Amount applied to principal = \$577.52 - \$54.70

= \$522.82 — *Principal*

5. Payment 1

With P = \$185,500, and $r = \frac{10.4\%}{12} = \frac{2.6}{3}\%$,

$I = \$185{,}500 \cdot \frac{2.6}{300} = \$1{,}607.67$, to the nearest cent *Interest*

Amount applied to principal = \$6,020.46 - 1,607.67

= \$4,412.79 *Principal*

Payment 2

With P = \$185,500.00 - 4,412.79 = \$181,087.21, and $r = \frac{2.6}{3}\%$,

$I = \$181{,}087.21 \cdot \frac{2.6}{300} = \$1{,}569.42$, to the nearest cent *Interest*

Amount applied to principal = \$6,020.46 - 1,569.42

= \$4,451.04 *Principal*

Payment 3

With P = \$181,087.21 - 4,451.04 = \$176,636.17, and $r = \frac{2.6}{3}\%$,

$I = \$176{,}636.17 \cdot \frac{2.6}{300} = \$1{,}530.85$ *Interest*

Amount applied to principal = \$6,020.46 - 1,530.85

= \$4,489.61 *Principal*

9. With monthly payments and a 2 year loan,

Number of payments = 12 • 2 = 24

Amount repaid = \$146.97 • 24 = \$3,527.28

Interest paid = \$3,527.28 - \$3,185.00

= \$342.28

13. With PV = \$375, $r = \frac{19.8\%}{12} = 1.65\%$, and n = 12,

$$PMT = \$375\left(\frac{0.0165}{1 - \frac{1}{(1 + 0.0165)^{12}}}\right)$$

$$= \$375\left(\frac{0.0165}{0.17830356}\right)$$

$$= \$375(0.09253881)$$

= \$34.70, to the nearest cent

17. With PV = \$41,000, $r = \frac{10.0\%}{2} = 5\%$, and n = 7,

$$PMT = \$41{,}000\left(\frac{0.05}{1 - \frac{1}{(1 + 0.05)^{7}}}\right)$$

$$= \$41{,}000\left(\frac{0.05}{0.28931867}\right)$$

$$= \$41{,}000(0.17281982)$$

= \$7,085.61

21. With PV = \$425, $r = \frac{12\%}{12} = 1\%$, and n = 36,

PMT = \$425 · 0.03321431 *PMT = PV • Table Factor*

= \$14.12 *Rounded to the nearest cent*

The Table Factor is found in A-8 under Column III, headed "Periodic payments necessary to pay off a loan". The Table Factor is the value of the factor $\left(\frac{r}{1 - \frac{1}{(1+r)^n}}\right)$ *for r = 1% and n = 36, rounded to eight decimal places.*

25. Based on the Rule of 78, the fractional part of payment one that is interest is $\frac{12}{78}$.

Thus, \$500.00 · $\frac{12}{78}$ = \$76.92 of the total interest is paid on the first payment.

29. Based on the Rule of 78, the fractional part of payment six that is interest is $\frac{7}{78}$.

Thus, \$1,250.80 · $\frac{7}{78}$ = \$112.25 of the total interest is paid on the sixth payment.

Exercise Set 9-4 Annuities and Sinking Funds

1. Payment 1

With PV = \$75,000, and $r = \frac{6\%}{12} = 0.5\%$

Interest portion = \$75,000 · 0.005 = \$375.00

Principal portion = \$832.65 - 375.00 = \$457.65

New principal = \$75,000.00 - 457.65 = \$74,542.35

Payment 2

Interest portion = \$74,542.35 · 0.005 = \$372.71

Principal portion = \$832.65 - 372.71 = \$459.94

New principal = \$74,542.35 - 459.94 = \$74,082.41

Payment 3

Interest portion = \$74,082.41 · 0.005 = \$370.41

Principal portion = \$832.65 - 370.41 = \$462.24

New principal = \$74,082.41 - 462.24 = \$73,620.17

5. Payment 1

With PV = \$31,950, and $r = \frac{11\%}{12} = \frac{11}{1200}$

Interest portion = \$31,950 · $\frac{11}{1200}$ = \$292.88

Principal portion = \$694.67 - 292.88 = \$401.79

New principal = \$31,950 - 401.79 = \$31,549.21

Payment 2

Interest portion = $\$31,549.21 \cdot \frac{11}{1200} = \$289.19$

Principal portion = $694.67 - 289.19 = $405.48

New principal = $31,549.21 - 405.48 = $31,143.73

Payment 3

Interest portion = $\$31,143.73 \cdot \frac{11}{1200} = \$285.47$

Principal portion = $694.67 - 285.48 = $409.20

New principal = $31,143.73 - 409.19 = $30,734.54

9. With PV = $41,375, $r = \frac{9\%}{12} = 0.75\%$, and n = 9 · 12 = 108

PMT = $41,375 · Table Factor — *PMT = PV • Table Factor*

= $41,375(0.01354291) — *Table A-5, Column III, in row n = 108.*

= $560.34, to the nearest cent

13. With PMT = $100, $r = \frac{8\%}{12} = 0.6666667\%$, and n = 60

FV = $100 · Table Factor — *FV = PMT • Table Factor*

= $100(73.4768564) — *Table A-4, Column V, in row n = 60.*

= $7,347.69, to the nearest cent

17. With PMT = $1,000, $r = \frac{9\%}{1} = 9\%$, and n = 25

FV = $1,000 · Table Factor — *FV = PMT • Table Factor*

= $1,000(84.7008963) — *Table A-38, Column V, in row n = 25.*

= $84,700.90, to the nearest cent

21. With PMT = $650, $r = \frac{7\%}{12} = 0.005833$, and n = 120

PV = $650(1 + 0.005833) Table Factor — *PV = PMT(1 + r) Table Factor*

= $653.791664(86.1263602) — *Table A-3, Column IV, in row n = 120.*

= $56,308.69

25. With FV = $10,000, $r = \frac{7\%}{12} = 0.005833$, and n = 5 · 12 = 60

PMT = $10,000 · Table Factor — *PMT = FV • Table Factor*

= $10,000(0.01396786) — *Table A-3, Column VI, in row n = 60.*

= $139.68

29. With FV = $3,200, $r = \frac{6.0\%}{4} = 1.5\%$, and n = 4 · 6 = 24

PMT = $3,200 · Table Factor — *PMT = FV • Table Factor*

= $3,200(0.03492410) — *Table A-14, Column VI, in row n = 24.*

= $111.76

PART C Chapter Nine Sample Test

1. Use the multistep method to compute the compound amounts for one year.

Principal	*Annual Interest Rate*	*Conversion Period*	
$2,500	10%	quarterly	1. *First quarter* 25625
			Second quarter 2564.0625
			Third quarter 2564.100625
			Fourth quarter 2564.00

In 2-4, for the indicated number of years, find (a) the compound amount and (b) the compound interest. Use the formula or the present value tables to compute each compound amount.

	Principal	*Annual Interest Rate*	*Number of Years*	*Conversion Period*	
2.	$5,000	12%	3	monthly	2. a. ________ b. ________
3.	$12,500	8.5%	5	semiannually	3. a. ________ b. ________
4.	$23,750	13.0%	2	quarterly	4. a. ________ b. ________

In 5 and 6, find the present value of each future value compound amount. Use the formula or the present value tables to compute each amount.

	Future Value	*Annual Interest Rate*	*Number of Years*	*Conversion Period*	
5.	$15,000	6%	4	monthly	5. ________
6.	$7,500	10%	8	quarterly	6. ________

7. Compute the interest and principal portions of each of the first two payments on the following amortized loan.

Present Value	*Annual Interest Rate*	*Monthly Payment*	*Term of Loan*
$20,000	11%	$434.85	5

7. *First payment* ______

Interest ______

Principal ______

Second payment ______

Interest ______

Principal ______

In 8 and 9, use the Rule of 78 to compute to the nearest cent the amount of the indicated payment that is interest.

	Number of Payments	*Payment Number*	*Total Interest*
8.	12	2	$832.20
9.	12	4	$315.65

8. ______

9. ______

10. Sharon Ganderholter has an ordinary annuity into which she makes monthly payments of $150. The annuity earns interest at an annual rate of 9% compounded quarterly.

a. What will be the value of the annuity in 4 years? 10. a. ______

b. How much of the amount in the annuity is interest? b. ______

11. High-Wall Builders will need $150,000 in six years to buy new equipment. They will make annual payments in a sinking fund that pays 8% annual interest compounded annually. Find the amount of the periodic payments.

11. ______

12. Sharon and Nancy Hardwicke bought a home that is financed with a $53,480 mortgage at 10% annual interest. The term of the loan is 30 years.

a. Find the amount of the monthly payments. 12. a. ______

b. Compute the total interest to the nearest dollar. b. ______

CHAPTER TEN

PERSONAL FINANCE

PART A Summary of Topics

OBJECTIVES

1. *Learn what is meant by a budget.*
2. *Use past expenditures as a guide for anticipated expenditures.*
3. *Learn how to analyze spending habits.*
4. *Learn how to budget for savings and contingencies.*
5. *Find out how to compare and adjust expenses to income.*
6. *Learn the parts of a checking account form.*
7. *Learn how to keep a check register.*
8. *Perform a bank reconciliation of a checking account.*
9. *Learn what is meant by a revolving charge account.*
10. *Know what is meant by the no-interest payment plan.*
11. *Compute interest based on the add-on-interest plan.*
12. *Compute interest on the interest-only plan.*

SECTION 10-1 Budgeting Personal Expenses

1. A budget is a listing of income and anticipated expenses for a specified time period. For most personal budgets, the time period is one month.

   A budget is one way of managing one's personal finances to safeguard against making financial obligations (also called "bills") beyond the ability to repay them. A little financial planning frequently reduces the pain associated with too much "going out", and too little "coming in".

2. Within most personal budgets it is possible to put expenses in three general categories.

   Category 1. Expenses that usually do not change from one time period to the next.

   Such as rent and long term loans.

   Category 2. Expenses that usually vary from one time period to the next.

   Such as food, utilities and entertainment.

   Category 3. Expenses that show a fixed change from one time period to the next because some event has happened.

   Such as the periodic payment of an insurance premium, a major car repair, or the starting of a new loan.

3. Two items that should be regularly included in a budget are:

   a. savings

   b. contingency fund

   One road to financial security is through a systematic and regular savings plan. A contingency fund will provide financial resources for unexpected bills and make regular savings a more realistic possibility.

4. When anticipated expenses exceed anticipated income, then a budget should be adjusted to bring the two figures into agreement. Two methods can be used to approach the problem:

   A. The specific amount method. The individual decides what items in the budget can be reasonably reduced, and by what amount.

   B. The percent method. All items in the budget that can be reasonably reduced are reduced by a percent that will bring anticipated expenses into agreement with anticipated income.

   The percent necessary to balance expenses and income can be calculated when the size of the difference is known.

$$\textit{Percent reduction} = \frac{\textit{Amount of reduction needed}}{\textit{Sum of amounts of items to be reduced}}$$

SECTION 10-2 Checking Accounts

1. Most checks contain at least the eight following items:

   a. The person that holds the account.

   b. The name of the bank that holds the money in the account.

   c. An account number.

   d. A bank number, called the American Banking Association number.

   e. A check number as indicated by the person that holds the account.

   f. The amount of money for which the check is written, both in numbers and words.

   g. The date the check was written.

   h. The name of the person or company to whom the check is being given.

2. A check register is a check by check accounting of the remaining amount of money in the account (called the balance).

   When a check is written, the amount of the check is subtracted from the balance. When a deposit is made, the amount of the deposit is added to the balance.

3. A bank reconciliation should be carried out each time an individual gets a statement of the account from the bank.

   The reconciliation means bringing into agreement the balances stated by A. the banking institution, and B. the individual's check register.

4. The following five steps can be followed to reconcile an individual's account with a bank statement.

   Step 1. The cancelled or cleared checks recorded on the bank statement should be checked (✓) on the check register.

   Step 2. The deposits indicated on the bank statement should be checked (✓) on the check register.

   Step 3. The checks listed on the check register that are not checked are outstanding. These should be listed in the appropriate column on the bank reconciliation worksheet.

   Step 4. The deposits listed on the check register that are not checked are in transit. These should be listed in the appropriate lines on the bank reconciliation worksheet.

Step 5. The bank reconciliation worksheet should be completed and the results compared with the stated balance on the individual's check register.

SECTION 10-3 Installment Purchases and Loans

1. A revolving charge account is one that extends to an individual consumer an open line of credit to some specified limit. The individual may then make purchases over some period of time, whereupon a summary of the account for that time period is sent to the consumer. Credit cards can be put into two categories.

   Category 1. International bank credit cards, such as Master Card and Visa.

   Category 2. Company credit cards, such as those issued by nationwide retail chains. Gasoline credit cards are included in this category.

2. The no-interest payment plan is one that permits the consumer to make a few fixed payments over a short period of time and pay no interest on the purchase.

   The frequently used phrase "90 days is the same as cash" means that the consumer can pay for the purchase within 90 days of the purchase date and not pay interest. The 30-60-90 plan is one in which the total purchase price is paid in three installments that are 30, 60 and 90 days after the purchase date.

3. The add-on interest plan is one in which the interest on the amount needed to purchase an item is added to the cost of the item. Regular payments are then made that pay the cost of the item and the interest on the loan, called a finance charge.

   The financing of large appliances and cars is frequently done this way. The cost of borrowing the money to buy the item is added to the purchase price to form a "package deal". When the final payment is made, the cost of the item and the finance charge are both paid-off at the same time.

4. The interest-only plan is one in which only the interest on the amount borrowed is regularly paid. No money is paid on the principal. With such a plan, only the periodic interest payments are made until some due date. On the due date the entire principal amount and the last interest payment must be made.

PART B Selected Solutions

Section 10-1 Budgeting Personal Expenses

1. a. $\text{Average} = \frac{75 + 75 + 138 + 138}{4} = \frac{426}{4} = 106.5$

   The average loan payment expense was $106.50.

   b. From July to August the loan payment expense jumped $63; probably the result of starting a new loan. Thus, the item belongs to Category 3.

   c. The anticipated expense is $138, a continuation of the expense shown in August and September.

5. a. $\text{Average} = \frac{330 + 330 + 330 + 330}{4} = \frac{1320}{4} = 330$

   The average rent expense was $330.

b. Rent is usually a constant figure, therefore the item belongs to Category 1.

c. The anticipated expense is \$330, a continuation of the expense shown June through September.

9. The percentage is \$305 of the total budget that is car expenses. The base of the percentage is the total budget \$2,600. If R is the rate of the total budget that is car expenses, then

$$R = \frac{305}{2600} \qquad R = \frac{P}{B}$$

= 0.117, to three decimal places

= 12%, to the nearest whole percent

13. The percentage of the total budget that is food is \$260. The base of the percentage is the total budget \$2600. If R is the rate of total budget that is food expense, then

$$R = \frac{260}{2600} \qquad R = \frac{P}{B}$$

= 0.10

= 10%

17. Contingency item = 0.10(1,815) *10% of \$1815*

= 181.50

Total anticipated budget = \$1,815.00 + 181.50

= \$1,996.50

21. Total anticipated budget = \$2500.00

Amount over income = \$2500 - 2250

= \$250 *Amount that needs to be cut.*

Total of items that can be cut = \$1335

Percent that each item must be cut = $\frac{250}{1335}$

= 0.187, to three decimal places

= 19%, to the nearest percent

Cut in auto expenses = 0.19 • 210 *19% of \$210*

= 39.90

Thus auto expenses should be cut about \$39.90.

25. From exercise 21, each item should be cut about 19%.

Cut in entertainment expenses = 0.19 • 275 *19% of \$275*

= 52.25

Thus entertainment expenses should be cut about \$52.25.

Exercise Set 10-2 Checking Accounts

1. Following is a display of the check register in Figure 10-A.

Check No.	Date	Check Issued To	Date of Dep	Amount of Deposit	Amount of Check	Balance $985.20
763	3/18	Westfall Apts.			$405.00	580.20
764	3/20	Best Products			42.78	537.42
765	3/21	Great Savings			71.00	466.42
766	3/25	GMAC			216.15	250.27
		Paycheck	3/26	535.73		786.00

5. Following is a display of the check register in Figure 10-E.

Check No.	Date	Check Issued To	Date of Dep	Amount of Deposit	✓	Amount of Check	Balance $3400.00
101	4/1	Morgan's Auto Repair			✓	$214.00	3186.00
102	4/4	Thrifty Drug Stores			✓	16.35	3169.65
		Deposit	4/10	$1200.00	✓		4369.65
103	4/12	Pacific Telephone			✓	18.50	4351.15
104	4/15	Pic N Pay Mkt.			✓	19.07	4332.08
105	4/20	Gladstone Apt's.			✓	500.00	3832.08
106	4/20	Cash			✓	100.00	3732.08
107	4/20	Sears			✓	22.26	3709.82
108	4/28	American Express				87.55	3622.27
		Deposit	4/28	425.00			4047.27
109	4/29	Chevron Oil				112.83	3934.44

To reconcile the above check register with the following bank statement a ✓ has been placed in the register next to the checks shown on the bank statement. A ✓ has also been placed next to the deposit shown on the bank statement.

Ending balance on the bank statement	$3709.82
Add the deposit of 4/29	+ 425.00
	$4134.82
Subtract the total of 108 and 109	- 200.38
	$3934.44
Enter the check book balance	$3934.44

Since the two figures agree perfectly, the account balances and no adjustments are necessary.

Exercise Set 10-3 Installment Purchases and Loans

1. The balance on the account is \$895.97

   Credit available = \$1,000.00 - 895.97 *Credit available = Credit limit - Balance*

   = \$104.03

5. Credit available = \$3,000.00 - 895.97

   = \$2,104.03 *Credit available = Credit limit - Balance*

9. With a balance of \$750 and a monthly interest rate of 1.65%

   Interest = \$750 • 0.0165 *Interest = (Balance)(Interest rate)*

   = \$12.375

   = \$12.38, to the nearest cent

13. Cost of item is \$1,000

    a. Down payment = $\$1,000 \cdot \frac{1}{4}$ *The terms are one-fourth down*

    = \$250

    Thus, a down payment of \$250 is required.

    b. Amount due = \$1,000 - 250

    = \$750

17. a. With a \$500 purchase price, a 10% annual interest rate, and a one year contract,

    Interest = \$500 • 0.10 • 1 $I = P \cdot R \cdot T$

    = \$50

    b. Contract price = \$500 + 50 *Contract price = Purchase price + Interest*

    = \$550

    c. Monthly payment = $\frac{\$550}{12}$ *Monthly payment* = $\frac{\textit{Contract price}}{12}$

    = \$45.83, to the nearest cent

21. a. With a \$1,200 purchase price, a 16% annual interest rate, and a $1\frac{1}{2}$ year contract,

    Interest = \$1,200 • 0.16 • 1.5 $I = P \cdot R \cdot T$

    = \$288

    b. Contract price = \$1200 + 288 *Contract price = Purchase price + Interest*

    = \$1,488

    c. Monthly payment = $\frac{\$1488}{18}$ *Monthly payment* = $\frac{\textit{Contract price}}{18}$

    = \$82.67, to the nearest cent

25. With the principal of $500, and an annual interest rate of 12%, and monthly payments,

Interest = $500 · 0.12

= $60 *The interest for one year.*

Monthly payment = $\frac{\$60}{12}$

= $5 *The interest for each month.*

29. With the principal of $3,600, and an annual interest rate of 13.5%, and annual payments,

Interest = $3,600 · 0.135

= $486

Since payments are made annually, the annual interest of $486 is the payment.

PART C Chapter Ten Sample Test

1. For each expense item in the personal budget:

   (i) Compute the average for the given months to the nearest dollar.

   (ii) Identify the category into which each item can be placed.

   (iii) State the anticipated expense for the next budget.

	Item	*August*	*September*	*October*	*Average*	*Category*	*Anticipated expense*
a.	Rent	$325	$325	$325	(i)______	(ii)______	(iii)________
b.	Food	210	178	226	(i)______	(ii)______	(iii)________
c.	Utilities	128	110	94	(i)______	(ii)______	(iii)________
d.	Car loan	248	248	248	(i)______	(ii)______	(iii)________
e.	Sears	43	37	98	(i)______	(ii)______	(iii)________

2. Compute the fractional part of a total budget of $1325 that belongs to the expense of each item. Write answers rounded to the nearest whole percent.

	Item	*Anticipated Expense*		
a.	Mortgage payment	$340	2. a.	____________
b.	Food	$160	b.	____________
c.	Utilities	$120	c.	____________
d.	Car loan	$238	d.	____________
e.	Revolving charge accounts	$106	e.	____________

3. Use the percent method to adjust the following budget to compare equitably with an income of $1,750. State to the nearest dollar the amount by which each item that can be reduced should be reduced. *For this budget, items a, b and c cannot be reduced.*

	Item	*Anticipated Expense*		
a.	Mortgage payment	$382	3. a.	____________
b.	Car loan	260	b.	____________
c.	Insurance premium	83	c.	____________
d.	Utilities	175	d.	____________
e.	Food	285	e.	____________
f.	Entertainment	250	f.	____________
g.	Contingency	200	g.	____________
h.	Revolving Credit payment	204	h.	____________

4. Complete a reconciliation of the check register balance (a) and the bank statement (b). Show all work.

a.	Check register balance		$692.65
b.	Bank statement balance		262.93
c.	Outstanding checks		
	No. 76	$ 10.56	
	No. 77	204.19	
	No. 78	73.75	
d.	Deposit made but not shown on bank statement		713.72
e.	Service Charge		4.50

5. Cynthia Thompson bought a portable VCR and camera package from Turner's TV and Camera Store for $1970. She used the no-interest plan, which included the following schedule:

10% payment at the time of purchase
25% payment 30 days after the date of purchase
35% payment 60 days after the date of purchase
30% payment 90 days after the date of purchase

a. Compute the payment at the time of purchase a. ____________
b. Compute the payment 30 days after the date of purchase b. ____________
c. Compute the payment 60 days after the date of purchase c. ____________
d. Compute the payment 90 days after the date of purchase d. ____________

6. Gary Morales bought a set of tires and rims at Metropolitan Tire Shop for $840. The Tire Shop offers a purchase plan that charges interest at an annual rate of 19.2% for each year of the contract. They have a 12 month, 18 month and 24 month contract.

a. Compute the monthly payments on the 12-month contract a. ____________
b. Compute the monthly payments on the 18-month contract b. ____________
c. Compute the monthly payments on the 24-month contract c. ____________

7. April Summers has a revolving charge card at a major department store with a credit limit of $2500. The store charges a 1.5% service charge on the beginning balance of the account each month. On this month's statement the beginning balance is $863.72.

a. Compute the credit available to April a. ____________
b. Compute the service charge on this month's statement b. ____________

8. James Palance bought a painting for $12,500. He made interest-only payments each quarter for two years at an annual interest rate of 12%. At the end of the two years the principal amount plus the last quarter's interest were due and payable.

a. Compute the amount of the quarterly interest payments. a. ____________

b. Compute the total amount due on the last payment b. ____________

CHAPTER ELEVEN

INVESTMENTS, REAL ESTATE, AND DEPRECIATION

PART A Summary of Topics

OBJECTIVES

1. *Learn what is a corporation and about the different kinds of corporations.*
2. *Learn what is a stock and about the three major classifications of stocks.*
3. *Learn the types of data given in a stock market quotation.*
4. *Learn how to compute the price-earnings ratio of a stock.*
5. *Learn the two methods used to compute the rate of return of a stock.*
6. *Learn how to compute the commission charged when stocks are bought or sold.*
7. *Learn what is a bond and what information is stated on a bond.*
8. *Learn the four general categories into which bonds are separated.*
9. *Learn how to compute the present value of a bond.*
10. *Learn the different types of real estate loans that are available.*
11. *Learn how to compute the principal and interest portion of the monthly payment on a real estate loan.*
12. *Learn what is meant by a loan origination fee.*
13. *Learn what is a loan or investment penalty.*
14. *Learn four penalties commonly assessed for early withdrawal from an investment fund.*
15. *Learn two penalties commonly assessed for early payoff of a loan.*
16. *Learn what is an investment discount.*
17. *Learn what is the rate of return of an investment.*
18. *Learn how to compute the effective rate of return of an investment.*
19. *Learn how to compute a base for an effective rate of return.*
20. *Learn how the cost of property as an expense is depreciated.*
21. *Learn how to depreciate property using the straight-line method.*
22. *Learn how to depreciate property using the double-declining balance method.*
23. *Learn how to depreciate property using the sum-of-the-year's-digits method.*
24. *Learn how to categorize property by type and class.*
25. *Learn how to depreciate real property.*

SECTION 11-1 Stocks

1. There are three basic categories into which ownership of a business can be separated.

   a. Sole proprietorships — *Businesses owned by one individual.*

   b. Partnerships — *Businesses owned by two or more individuals.*

   c. Corporations — *Businesses owned by many individuals.*

2. The ownership of a corporation is divided into units called stocks, and the owners are called stockholders. There are three general classifications of corporations.

   a. Private corporations — *Stock is not usually for public sale.*

   b. Public corporations — *Stock is widely held and publicly traded.*

   c. Nonprofit corporations — *Stock may not be issued.*

3. A stock is a certificate that states that the holder of the certificate is a part owner of the corporation named on the certificate. Three major classifications of stock are:

   a. Common — *The rate of dividends can vary.*

   b. Preferred — *The dividends cannot be changed.*

   c. Cumulative preferred — *The dividends can accumulate if they cannot be paid when due.*

4. The price-earnings ratio is found by dividing the current price per share by the current net profit per share.

$$\text{P-E ratio} = \frac{\text{current price per share}}{\text{current net profit per share}}$$

5. The rate of return of a stock can be calculated in two ways:

   a. Based on the cost of the stock to the stockholder

$$\text{Rate of return} = \frac{\text{current dividend per share}}{\text{cost of stock to stockholder}}$$

   b. Based on the current selling price of the stock

$$\text{Rate of return} = \frac{\text{current dividend per share}}{\text{current selling price per share}}$$

6. When stocks are bought or sold through a brokerage firm, the agent charges a commission, and the rate of commission is usually a percent of the purchase price.

   The rate of commission is usually more for odd lots than for even lots. Even lots are multiples of 100, and odd lots are not.

SECTION 11-2 Bonds

1. A bond is a written promise wherein the borrower promises to pay to the holder of the bond:

   a. the face value of the bond at the maturity date, and

   b. interest at an annual rate stated on the bond.

2. The following details are given on a bond:

   a. the issuer of the bond — *The corporation or government agency that is selling the bond.*

   b. The face value of the bond — *The amount to be paid to the holder of the bond on the date of maturity.*

   c. The annual interest rate — *The rate at which interest is to be paid.*

d. The maturity date — *The date on which the face value of the bond is to be paid.*

e. The interest dates — *The dates on which the interest is to be paid, usually semiannually.*

3. Four general classifications of bonds

   a. Municipal bonds — *Bonds sold by a municipality, such as city, county, or school district.*

   b. Corporation bonds — *Bonds sold by a corporation.*

   c. Secured bonds — *Bonds that are backed by some form of property.*

   d. Unsecured bonds — *Bonds that are not backed by some form of property.*

4. The present value of a bond can be calculated by using the techniques studied in Chapter 9. The technique requires four steps.

Step 1. Compute the amount of the periodic interest payments.

Use the equation,

$I = P \cdot r$

P is the face value of the bond

r is the periodic interest rate

If interest is paid annually, then r = R, the interest rate stated on the bond. If interest is paid semiannually, then $r = \frac{R}{2}$, that is, one-half the annual interest rate.

Step 2. Compute the present value of the total interest to be received for the remaining life of the bond. Now the current market annual interest rate is used, and the periodic interest rate used in the formula is

$$r = \frac{\text{current market interest rate}}{\text{number of times interest is paid each year}}$$

Use the equation,

$$PV = PMT\left(\frac{1 - \frac{1}{(1 + r)^n}}{r}\right)$$

PMT = periodic interest payment

r = periodic interest rate

n = number of remaining interest periods

or, $PV = PMT \cdot \text{Table Factor}$

The Table Factor is found in the Present Value Table for r% and under Column IV.

Step 3. Find the present value of the face value of the bond. Again, use R = the current market annual interest rate to determine the r in the following formula. Use the equation,

$$PV = FV\left(\frac{1}{(1 + r)^n}\right)$$

FV = face value of the bond

r = periodic interest rate

n = number of remaining periods

or, $PV = FV \cdot \text{Table Factor}$

The Table Factor is found in the Present Value Table for r% and under Column II.

Step 4. Add the PV's obtained in Steps 2 and 3.

These steps determine the amount of money that is needed to invest in a similar compounding interest investment to return the amount to be received by the bond (face value of bond + interest). This is the reason that the current market interest rate is used, and not the interest rate stated on the bond.

SECTION 11-3 Real Estate

1. Some of the more common types of real estate loans are:

   a. Conventional - excellent credit rating, large downpayment, fixed interest rate, and fixed periodic payments.

   b. Government-assisted - government guarantee on payments, low downpayment, and lower interest rate.

   c. Variable-interest-rate - interest rate can go up or down, the periodic payments can change, and a possible "balloon" payment at the end of the loan period.

2. A graduated-interest-rate real estate loan provides a schedule that allows the interest rate to increase over some specified time period.

   Such a loan may help first time home buyers afford mortgage payments. The theory is that a family unit's income will increase with time, thus making larger payments possible in the future. Thus, payments increase as income increases.

3. A graduated-payment real estate loan provides a schedule that allows the payments to increase over some specified time period.

   The theory to such a loan package is similar to the graduated-interest-rate loan. In this case the interest may remain the same, but the payments will increase. A disadvantage to this plan is that the initial payments may not be large enough to pay the interest portion, and as a consequence the unpaid interest is added to the principal amount owed on the house. In such cases a home buyer may owe more on the house after one or two years payments than the original principal on the mortgage loan.

4. A balloon (or call) real estate loan provides for a lump sum payment at the end of the loan period.

5. Many real estate loans require a loan origination fee. Such a fee is frequently referred to as points. The term "points" is used in two ways:

   a. the points are the rate used to compute the loan fee.

   b. the points are the stated amount of the fee.

SECTION 11-4 Loan and Investment Penalties

1. A loan or investment penalty may be described as a fee, charge, or fine for the reduction of the time period of a loan or investment.

   The penalty can be levied for paying a loan off in less time than the full term of the loan. Such a penalty is called prepayment. Notice that a revolving charge account has no such penalty.

   The penalty can also be levied for taking money out of an investment fund that has a specified time period. Such a penalty is called "early withdrawal".

2. Some of the ways in which a penalty may be assessed for early withdrawal from an investment account are:

   a. Lose all the interest earned by the investment.

   b. Reduce the rate at which interest is earned by the investment.

   c. Lose the interest earned for some specified time period.

   d. Lose a specified percent of the interest earned.

3. Some of the ways in which a penalty may be assessed for early payoff of a loan are:

   a. Pay a portion of the interest that would have been paid for a specified time period.

   b. Pay a percent of the remaining balance at the time of payoff.

SECTION 11-5 Discounts and Effective Rate of Return

1. The face or par value of an investment is the stated value.

   An investment is discounted when it is sold for an amount less than the face value.

   Amount of discount = face value • discount rate

   Current selling price = face value - discount

2. The effective rate of return on an investment is the percent obtained by dividing the annual income of the investment by the cost of the investment.

$$\text{Effective rate of return} = \frac{\text{Annual income from the investment}}{\text{Cost of the investment}}$$

3. The following five steps can be used to find the amount of discount necessary to achieve an effective rate of return for an investment:

   Step 1. Find the amount of the periodic interest payments.

   Use the interest rate stated on the investment document.

   Step 2. Find the present value of the total interest to be received.

   Use the current market interest rate.

   Step 3. Find the present value of the face value of the note.

   Again, use the current market interest rate.

   Step 4. Add the present values obtained in Steps 2 and 3.

   Step 5. Subtract the total present value of Step 4 from the face value of the note to find the amount of discount.

   These five steps basically determine how much money in the present is needed to accumulate an amount in the future based on the current market interest rate. The investment that must be discounted is earning interest at a lower rate, thus it will earn a lesser amount. The difference in what it should earn and what it will earn is the amount of discount needed to achieve the desired rate of return.

SECTION 11-6 Depreciation

1. Depreciation is the distribution of a portion of the cost of property as an expense to future income periods.

   By depreciating the cost of a piece of property over the useful life of the property, all of the accounting periods during this time period will reflect the cost of that item, and not just the accounting period in which the item was purchased.

2. The straight-line method of depreciation distributes equal amounts of the wearing value of an item (or property) to each year of the useful life of the item.

   D = the annual depreciation

   C = the original cost of the item

   S = the salvage (or scrap) value of the item

   N = the number of years of useful life for the item

$$D = \frac{C - S}{N}$$

3. Accumulated depreciation is the sum of the depreciations over the previous years. Book value of an item is the difference between the original cost and the accumulated depreciation.

   Book value = original cost - accumulated depreciation

   Book value (also called "Blue Book") is commonly used in conversations related to cars and trucks. To the consumer, the Blue Book on a car is the present value of the car. As seen by the equation that defines book value, the figure on a given car is the difference between the original cost and the accumulated depreciation.

4. The double-declining balance method of depreciation is a method whereby a larger amount of depreciation is realized in the earlier years of the life of an item than in the later years. For each year (except possibly the last) the depreciation is the product of the depreciable base (= original cost - accumulated depreciation) and the factor 2/N, where N is the number of years of useful life.

5. The sum-of-the year's- digits method of depreciation is another way in which larger amounts of depreciation can be claimed for the earlier years of useful life of an item. If N is the number of years of useful life, then

$$N + (N - 1) + (N - 2) + \ldots + 2 + 1$$

   is the sum of the digits 1 through N. This sum can be computed with the formula,

$$\text{Sum of year's digits} = \frac{N(N + 1)}{2}$$

   For the sum-of-the-year's-digits method of depreciation, the depreciable base is the same for each year.

   depreciable base = original cost - scrap value

SECTION 11-7 Accelerated Cost Recovery System (ACRS)

1. For income tax purposes, the depreciation of property is governed by the Accelerated Cost Recovery System, or simply ACRS. This system was adopted by the federal government on January 1, 1981.

2. Property is generally classified into two types, personal and real.

   a. Personal property is property that is not attached to real property.

      In a way, think of personal property as belonging to a person and can be taken anywhere with that person.

   b. Real property is land or anything that is attached to the land.

      Buildings, driveways, fences, and signs are examples of real property.

3. For depreciation purposes, the IRS places all personal property into two groups: three-year property and five-year property. The IRS has issued specific guidelines and charts for determining in which class a property belongs. The rates of depreciation for three- and five-year property are in the following tables:

Schedule for three-year property	
Year	ACRS Percent
1	25%
2	38%
3	37%

Schedule for five-year property	
Year	ACRS Percent
1	15%
2	22%
3	21%
4	21%
5	21%

4. To use an ACRS depreciation table for real property, the month the property was purchased must also be known. A depreciation schedule (issued by the IRS) can then be used to determine the percent depreciation that can be used for that year.

PART B Selected Solutions

Exercise Set 11-1 Stocks

1. For the stock listed Sperry in Figure 11-A:

   a. the highest selling price per share for the year is \$46 $\frac{1}{2}$

   b. the lowest selling price per share for the year is \$21 $\frac{5}{8}$

   c. the current dividend per share is \$1.92

   d. the yield percent is 4.2%

      The yield percent is the current dividend divided by current selling price.

   e. the price-earnings ratio is 16

   f. the number of shares sold was 155,200

   g. the highest selling price per share was \$45 $\frac{7}{8}$

   h. the lowest selling price per share was \$45 $\frac{3}{8}$

   i. the closing price per share was \$45 $\frac{3}{4}$

j. the net change in the price per share was $+\frac{1}{4}$

The stock increased $\$\frac{1}{4}$ (that is, 25 cents) this day.

5. Faith owns 500 of the 25,000 shares, thus

$$\text{Percent ownership} = \frac{500}{25{,}000} \qquad R = \frac{P}{B}$$
$$= 0.02$$
$$= 2\%$$

9. a. With 200,000 shares and \$785,000 net profit last quarter,

$$\text{Earnings per share} = \frac{\$785{,}000}{200{,}000}$$
$$= \$3.925 \qquad \textit{About \$3.93 per share}$$

b. With a current price per share of \$35, and a current net profit per share of \$3.925,

$$\text{Price-Earnings ratio} = \frac{\$35}{\$3.925}$$
$$= 8.9\text{, to one decimal place}$$

13. a. With cost per share = \$29, and current dividend = \$2.60

$$\text{Rate of return} = \frac{\$2.60}{\$29} \qquad \frac{\textit{current dividend}}{\textit{cost per share}}$$
$$= 0.090\text{, to three decimal places}$$
$$= 9.0\%\text{, to the nearest tenth of a percent}$$

b. With current selling price = $\$42\frac{1}{4}$, and current dividend = \$2.60

$$\text{Rate of return} = \frac{\$2.60}{\$42.25} \qquad \frac{\textit{current dividend}}{\textit{current selling price}}$$
$$= 0.062\text{, to three decimal places}$$
$$= 6.2\%\text{, to the nearest tenth of a percent}$$

17. With 200 shares, \$12.75 price per share, and 2.5% rate of commission,

$$\text{Commission} = 200(\$12.75)(0.025) \qquad 2.5\% = 0.025$$
$$= \$63.75$$

Exercise Set 11-2 Bonds

1. $10\frac{3}{4}$ S 94 means the interest rate on the bonds is $10\frac{3}{4}$ %, and the due date is in 1997.

15 means \$15,000 in bonds were traded during the week

$74\frac{1}{2}$ means the highest price for the week was \$745.00

$74\frac{1}{2}$ means the lowest price for the week was \$745.00

$74\frac{1}{2}$ means the last price for the week was \$745.00

$-\frac{5}{8}$ means the price per bond dropped $\frac{5}{8}$ of \$10.00 = \$6.25 from the previous week's close.

5. Interest on bonds: 5(0.0925)(1,000) = \$462.50 *one bond*

Interest on bonds:	5(0.0925)(1,000) =	\$462.50	*one bond*
	12(462.50) =	\$5,550.00	*twelve bonds*
Face value of bonds:	12(1,000) =	12,000.00	
Total return on bonds:		\$17,550.00	
Cost of bonds:	12(885) =	10,620.00	
Amount of money returned to holder:		\$ 6,930.00	

9. *Step 1.* *Compute the amount of the periodic interest payments.*

With R = 9% and semiannual interest payments, r = 4.5%.

I = 10,000 • 0.045 = 450 *\$450 interest each period*

Step 2. *Compute the present value of the total interest to be received for the remaining life of the bond.*

Now use the current market interest rate R = 10%, and r = 5%. With four years remaining in the life of the bonds and semiannual compounding n = 4 • 2 = 8. The Table Factor is found in table A-30, Column IV and n = 8.

PV = 450 • 6.46321275

= 2908.44574

= \$2,908.45, to the nearest cent

Notice, 8 interest payments of \$450 each yields \$3600. But the present value of the \$3600 is only \$2,908.45.

Step 3. *Find the present value of the face value of the bond.*

With r = 5% and n = 8, the Table Factor is found in table A-30, Column II.

PV = 10,000 • 0.67683936

= 6768.3936

= \$6,768.39, to the nearest cent

Notice, the holder of the bonds will get \$10,000 in four years, but the present value of that amount is \$6,768.39, since that much money in a semiannual compound interest investment at the current market interest rate of 10% will build to \$10,000 in four years.

Step 4. *Add the PV's of Steps 2 and 3.*

Present value of interest payments:	\$2,908.45
Present value of face value of bonds:	6,768.39
Total present value:	\$9,676.84

Thus, \$9,676.84 deposited today in an investment fund that yields 10% annual interest compounded semiannually will yield in four years the same amount that the holder of these bonds will get in the same time period.

Exercise Set 11-3 Real Estate

1. a. With PV = \$75,000, $r = \frac{10\%}{12} = 0.8333333\%$, and $n = 20 \cdot 12 = 240$ (20 year loan and monthly payments), the Table Factor is found in table A-6 and Column III.

   $PMT = 75{,}000 \cdot 0.00965022$ *PMT = PV • Table Factor*

   $= 723.7665$

   $=$ \$723.77, to the nearest cent

   b. Total amount paid $= 723.77 \cdot 240$ *(Periodic payment)(Number of payments)*

   $= 173{,}704.80$

   Interest paid = \$173,704.80 - 75,000.00

   = \$98,704.80   *Total interest paid*

5. a. With PV = \$195,650, $r = \frac{13\%}{12} = 1.083333$, and $n = 40 \cdot 12 = 480$ (40 year loan and monthly payments), the Table Factor is found in table A-9 and Column III.

   $PMT = 195{,}650 \cdot 0.01089514$

   $= 2131.63414$

   $=$ \$2,131.63, to the nearest cent

   b. Total amount paid $= 2{,}131.63 \cdot 480$ *(Periodic payment)(Number of payments)*

   $= 1{,}023{,}182.40$

   Interest paid = \$1,023,182.40 - 195,650.00

   = \$827,532.40   *Total interest paid*

9. The Principal Amount of the Loan is P for the first payment.

   a. With P = \$46,000 and $r = \frac{13.75\%}{12}$

   $I = 46{,}000 \cdot \frac{13.75}{1200}$ $\frac{13.75\%}{12} = \frac{13.75}{1200}$

   $= 527.083333$

   $=$ \$527.08, to the nearest cent   *Interest on first payment*

   b. With PMT = \$535.95 and I = \$527.08,

   Amount applied to principal = \$535.95 - 527.08

   = \$8.87

13. With P = \$45,000 and rate at which fee is charged = $3\frac{1}{4}\%$,

Loan origination fee = 45,000 • 0.0325 *$3\frac{1}{4}\% = 0.0325$*

= \$1,462.50

17. With P = \$102,600 and rate at which fee is charged = 3%,

Loan origination fee = 102,600 • 0.03

= \$3,078.00

<u>Exercise Set 11-4</u> <u>Loan and Investment Penalties</u>

1. The rate with which the penalty is assessed is the product of 10% and the number of months early the money is withdrawn.

   R = 10% • 3 = 30%

   The penalty is the percentage and the base is the amount of the scheduled interest.

   Penalty = \$750 • 0.30 *P = B • R*

   = \$225.00

   The 3 month early withdrawal will cost the investor \$225.

5. The penalty is the difference between the interest that would have been earned at 14% and the interest earned at the penalty rate of 10%.

   a. *<u>Step 1</u>. Compute the interest at 14% compounded semiannually for one year.*

   With PV = \$85,000, $r = \frac{14\%}{2} = 7\%$, and n = 2, the Table Factor is found in table A-34 and Column I.

   FV = 85,000 • 1.14490000 *FV = PV • Table Factor*

   = 97,316.5

   Interest = \$97,316.50 - 85,000.00

   = \$12,316.50 *The interest that would have been earned*

   <u>Step 2</u>. Compute the interest at 10% compounded semiannually for one year.

   With PV = \$85,000, $r = \frac{10\%}{2} = 5\%$, and n = 2, the Table Factor is found in A-30 and Column I.

   FV = 85,000 • 1.10250000 *FV = PV • Table Factor*

   = 93,712.50

   Interest = \$93,712.50 - 85,000.00

   = \$8,712.50 *The interest that is actually earned.*

   <u>Step 3</u>. Compute the difference in the interests obtained in Steps 1 and 2.

   Penalty = \$12,316.50 - 8,712.50

   = \$3,604.00

b. Step 1. With PV = $85,000, r = 7% and n = 5 (that is, five compounding periods),

FV = 85,000 • 1.40255173 *Table A-34 and Column I*

= 119,216.90, to the nearest cent

Interest = $119,216.90 - 85,000.00

= $34,216.90 *The interest that would have been earned*

Step 2. With PV = $85,000, r = 5% and n = 5,

FV = 85,000 • 1.27628156 *Table A-30 and Column I*

= 108,483.93, to the nearest cent

Interest = $108,483.93 - 85,000.00

= $23,483.93 *The interest that is actually earned*

Step 3. Penalty = $34,216.90 - 23,483.93

= $10,732.97

9. With P = $7,632, R = 10% and T = $\frac{4}{12} = \frac{1}{3}$ of a year,

I = 7,632 • 0.10 • $\frac{1}{3}$ *I = P • R • T*

= 254.4

The penalty is $254.40

Exercise Set 11-5 Discounts and Effective Rate of Return

1. a. With a face value of $1,000 and discount rate of $9\frac{1}{4}$ %,

discount = 1,000 • 0.0925 *discount = (face value)(discount rate)*

= 92.5

The discount is $92.50

b. Current selling price = $1,000.00 - 92.50 *current selling price = face value - discount*

= $907.50

5. a. With a face value of $12,360 and discount rate of $5\frac{1}{4}$ %,

discount = 12,360 • 0.0525 *discount = (face value)(discount rate)*

= 648.9

The discount is $648.90

b. Current selling price = $12,360.00 - 648.90 *Current selling price = face value - discount*

= $11,711.10

9. With an annual return of $85 and a cost of $930,

Effective rate of return $= \frac{85}{930}$ *Effective rate of return* $= \frac{\text{annual income}}{\text{cost}}$

= 0.0914, to four decimal places

= 9.14%

13. With an annual return of 4($157.28) = $629.12 and a cost of $3,200,

Effective rate of return $= \frac{629.12}{3200}$

= 0.1966

= 19.66%

17. With an annual return of 4($406.25) = $1,625 and a cost of $16,250,

Effective rate of return $= \frac{1,625}{16,250}$

= 0.1000

= 10.00%

21. *Step 1.* *Find the amount of the periodic interest payments.*

With P = $1,000 and r = 9% (the annual interest rate stated on the investment),

I = 1,000 · 0.09 *Interest is compounded annually*

= 90 *The periodic interest is $90.00*

Step 2. *Find the present value of the total interest to be received.*

With PMT = $90, r = 10% (the desired effective rate of return), and n = 2 (the remaining life of the investment),

PV = 90 · 1.73553720 *Table A-40 and Column IV*

= 156.19835

Step 3. *Find the present value of the face value of the note.*

With FV = $1,000, r = 10%, and n = 2,

PV = 1,000 · 0.82644628 *Table A-40 and Column II*

= 826.44628

Step 4. *Add the present values obtained in Steps 2 and 3.*

Total PV = 156.19835 + 826.44628

= 982.64, to two decimal places.

Step 5. *Subtract the total PV from the face value of the note.*

Discount = $1,000.00 - 982.64

= $17.36, the amount of the discount

25. Step 1. With P = \$5,000 and r = 6%,

$$I = 5{,}000 \cdot 0.06 = 300$$

The periodic interest payments are \$300.

Step 2. With PMT = \$300, r = 13%, and n = 7,

$$PV = 300 \cdot 4.42261043$$ *Table A-43 and Column IV*

$$= 1326.78313$$

The present value of the interest to be received is about \$1,326.78.

Step 3. With FV = \$5,000, r = 13%, and n = 7,

$$PV = 5{,}000 \cdot 0.42506064$$ *Table A-43 and Column II*

$$= 2125.30320$$

The present value of the face value of the note is about \$2,125.30

Step 4. Total PV = 1326.78313 + 2125.30320

= 3,452.08633

Step 5. Discount = 5,000 - 3,452.08633

= 1,547.91367

= \$1,547.91, to the nearest cent

Exercise Set 11-6 Depreciation

1. With C = \$7,500, S = \$500, and N = 3 years,

$$D = \frac{\$7{,}500 - 500}{3} = \$2{,}333.33$$

End of Year	Depreciable Base	Depreciation Expense	Accumulated Depreciation	Book Value
1	\$7,500.00	\$2,333.33	\$2,333.33	\$5,166.67
2	5,166.67	2,333.33	4,666.66	2,833.34
3	2,833.34	2,333.33	7,000.00	500.00

3. With C = \$150,000, S = \$15,000, and N = 15 years,

$$D = \frac{\$150{,}000 - 15{,}000}{15} = \$9{,}000$$

End of Year	Depreciable Base	Depreciation Expense	Accumulated Depreciation	Book Value
1	$150,000	$9,000	$ 9,000	$141,000
2	141,000	9,000	18,000	132,000
3	132,000	9,000	27,000	123,000
4	123,000	9,000	36,000	114,000
5	114,000	9,000	45,000	105,000
6	105,000	9,000	54,000	96,000
7	96,000	9,000	63,000	87,000
8	87,000	9,000	72,000	78,000
9	78,000	9,000	81,000	69,000
10	69,000	9,000	90,000	60,000
11	60,000	9,000	99,000	51,000
12	51,000	9,000	108,000	42,000
13	42,000	9,000	117,000	33,000
14	33,000	9,000	126,000	24,000
15	24,000	9,000	135,000	15,000

7. For the delivery van N = 5. Thus $\frac{2}{N} = \frac{2}{5}$

End of Year	Depreciable Base	Depreciation Expense	Accumulated Depreciation	Book Value
1	$26,500.00	$10,600.00	$10,600.00	$15,900.00
2	15,900.00	6,360.00	16,960.00	9,540.00
3	9,540.00	3,816.00	20,776.00	5,724.00
4	5,724.00	2,289.60	23,065.60	3,434.40
5	3,434.40	1,373.76	24,439.36	2,060.64

Notice that on year 5 a depreciation of $434.40 was taken to yield a Book Value that is the same as the Scrap Value (= $3,000).

9. For the automobile N = 3, thus the sum-of-the-year's-digits is 6.

End of Year	Depreciable Base	Depreciation Expense	Accumulated Depreciation	Book Value
1	$7,000.00	$3,500.00	$3,500.00	$4,000.00
2	7,000.00	2,333.33	5,833.33	1,666.67
3	7,000.00	1,166.67	7,000.00	500.00

11. For the building, $N = 15$. Thus, the sum-of-the-year's-digits = 120. The depreciable base for all 15 years is the same, that is, $150,000 - 15,000 = $135,000.

For year one, depreciation expense = $\$135,000 \cdot \frac{15}{120} = \$16,875$.

End of Year	Depreciable Base	Depreciation Expense	Accumulated Depreciation	Book Value
1	$135,000	$16,875	$ 16,875	$133,125
2	135,000	15,750	32,625	117,375
3	135,000	14,625	47,250	102,750
4	135,000	13,500	60,750	89,250
5	135,000	12,375	73,125	76,875
6	135,000	11,250	84,375	65,625
7	135,000	10,125	94,500	55,500
8	135,000	9,000	103,500	46,500
9	135,000	7,875	111,375	38,625
10	135,000	6,750	118,125	31,875
11	135,000	5,625	123,750	26,250
12	135,000	4,500	128,250	21,750
13	135,000	3,375	131,625	18,375
14	135,000	2,250	133,875	16,125
15	135,000	1,125	135,000	15,000

Notice that the Book Value after the fifteenth year is the Scrap Value $15,000.

Exercise Set 11-7 Accelerated Cost Recovery System (ACRS)

1. a. For the first year of a three-year personal property item, the depreciation rate is 25%. The original cost is $35,400.

   First year depreciation = $35,400 · 0.25

   = $8,850.

   b. For the third year of a three-year personal property item, the depreciation rate is 37%. The original cost is $35,400.

   Third year depreciation = $35,400 · 0.37

   = $13,098.

5. a. For the first year of a five-year personal property item, the depreciation rate is 15%. The original cost is $27,650.

   First year depreciation = $27,650 · 0.15

   = $4,147.50

b. For the fourth year of a five-year personal property item, the depreciation rate is 21%. The original cost is $27,650.

$$\text{Fourth year depreciation} = \$27,650 \cdot 0.21$$
$$= \$5,806.50$$

9. a. For real property bought in the third month, the depreciation rate for the first year is 10%. The original cost is $36,500.

$$\text{First year depreciation} = \$36,500 \cdot 0.10$$
$$= \$3,650$$

b. For real property bought in the third month, the depreciation rate for the sixth year is 6%. The original cost is $36,500.

$$\text{Sixth year depreciation} = \$36,500 \cdot 0.06$$
$$= \$2,190$$

PART C Chapter Eleven Sample Test

1. Hartnell and Dunsmire Corporation reported a net profit last quarter of $462,870. The Corporation has 125,000 shares outstanding, and the current selling price per share is $43 $\frac{1}{2}$.

   a. To the nearest cent, find the earnings per share for the last quarter. 1. a. ________

   b. To one decimal place, find the price-earnings ratio for last quarter. b. ________

2. Barry Michaels bought shares of Tin Can Corporation for $11 $\frac{3}{8}$ per share. The stock is now selling for $19 per share. The current dividend was $3.98 per share.

   a. Find the rate of return based on the cost of the stock to Michael to the nearest whole percent. 2. a. ________

   b. Find the rate of return based on the current selling price of the stock to the nearest whole percent. b. ________

3. Ken Wiznouski bought five municipal bonds for $925 each. The face value of each bond was $1,000 and the annual interest rate on the bonds was 7 $\frac{3}{4}$ %. The remaining life of the bonds was six years. Compute the amount of money returned to Ken above the original investment. 3. ________

4. The total face value of some bonds is $5,000. The stated interest rate on the bonds is 7%, semiannually, and the remaining life of the bonds is five years. The current market interest rate for similar investments is 11%. Compute the present value of the bonds. 4. ________

5. Anne and Noah Lockhart just bought a new home. They obtained a 30 year $63,450 amortized loan to buy it. The annual interest rate is 11% with $604.25 a month in payments.

   a. To the nearest cent, compute the amount of the first payment that is interest. 5. a. ________

   b. Compute the total interest paid if the loan is fully amortized. b. ________

6. Compute the amount of the fee on a loan of $15,000 if the points are 2 $\frac{3}{4}$. 6. ________

7. The remaining principal on a loan is $7,400. The annual interest rate is 14%, and the penalty for early payment is 6 months interest based on 14% and the remaining principal. Compute the penalty for early payment of this loan. 7. ______

8. Deborah White purchased a second mortgage of $22,500 on a piece of real estate. She gets quarterly returns of $855 each on the investment. To the nearest tenth of a percent, find the effective rate of return on the investment. 8. ______

9. A lease on a commercial building has stated value of $74,800. The lease is currently being sold at a 12% discount. Find the amount of the discount. 9. ______

10. Sonja Michaels bought a new truck for her horse ranch. The cost of the truck new was $12,800, the useful life is five years, and the scrap value is $2,300. Find the depreciation the first year using:

    a. the straight-line method 10. a. ______

    b. the double-declining method b. ______

    c. the sum-of-the-year's-digits method c. ______

11. Use the ACRS Depreciation Schedule for Real Property to find the depreciation for the 3rd year of a rental house that cost $43,500, if the house was bought in August. The price $43,500 is for the house only and does not include the cost of the land on which it is built. 11. ______

CHAPTER TWELVE

INSURANCE AND TAXES

PART A Summary of Topics

OBJECTIVES

1. *Learn what is meant by insurance.*
2. *Calculate the annual premiums for a term life insurance policy.*
3. *Calculate the annual premiums for an endowment life insurance policy.*
4. *Learn what is property insurance.*
5. *Study the factors that affect the premiums on automobile insurance.*
6. *Study the factors that affect the premiums on structure insurance.*
7. *Learn what is meant by co-insurance.*
8. *Learn what is meant by property tax.*
9. *Study some of the factors that affect the amount of property tax.*
10. *Compute the property tax for a given assessed value and tax rate.*
11. *Learn what is meant by an income tax and who must pay the tax.*
12. *Study a recommended procedure for computing a personal income tax.*
13. *Compute a personal income tax using tax tables.*

SECTION 12-1 Life Insurance

1. Insurance is a fund into which many individuals contribute a small amount of money. The purpose of the fund is to cover, or pay, the large unpredictable losses of a few of the contributing individuals.

    In the case of life insurance the loss experienced by the death of an individual is frequently the income the individual would have earned. Thus, someone that contributes to the income of a family unit will carry insurance on her or his life. Thus, should the individual die, the insurance will cover the income that the individual would have earned had she or he not died.

2. Life insurance is a contract (policy) that is used to provide money upon the death of the insured (the person whose life is insured) to the insured's beneficiaries (the recipients of the money as specified by the contract).

    a. *A term insurance policy is one that insures the life of the insured for a specified period. At the end of the period, the contract terminates. With such a policy there are no residual money benefits beyond the termination of the policy.*

    b. *An endowment life insurance policy is one that insures the life of the insured for a specified period and also provides a savings or accumulation fund. The accumulation fund earns interest (called dividends). With such a policy there are residual money benefits beyond the termination of the policy.*

3. To use Table 12-1 to compute the annual premium for a life insurance policy, follow the procedure described in Steps 1, 2 and 3.

    Step 1. Divide the amount of the face value of the policy by 1,000.

Step 2. Locate the age of the insured individual in the "Age of Insured" column and the corresponding rate in the appropriate "Term Life" or "Endowment Life" column.

Step 3. Multiply the quotient obtained in Step 1 by the rate obtained in Step 2.

The product obtained in Step 3 is the annual premium for the policy.

SECTION 12-2 Property Insurance

1. Property insurance is insurance that is bought by an individual or business to cover the loss of some property or equipment.

   Property insurance is also known as fire and casualty insurance.

2. Some of the components that are included in an automobile or truck insurance policy are the following:

   a. Bodily injury - *Pays the medical expenses of others resulting from the insured vehicle. This coverage does not include injuries to the individual who has insured the vehicle.*

   b. Property damage - *Pays the cost of repairing or replacing the property of others damaged by the insured vehicle.*

   c. Medical payments - *Pays the medical expenses of the insured and the insured's family when injuries are related to the insured vehicle.*

   d. Uninsured motorist - *Pays the expenses of the insured party when the damages are caused by another motorist who has no insurance.*

3. Home, buildings, and personal possessions may also be covered by insurance. As in the case of vehicle insurance, the property can be insured against loss and/or damage.

   Several factors influence the cost of insuring such items of property. With regards to buildings, some of the more obvious factors are the insuring company, the location of the property, the age and type of construction of the property, and the amount of deductible stated in the policy.

4. A co-insurance clause in an insurance policy on a structure means that the insurance company will pay less than the amount of damages that the structure might experience. With a co-insurance clause:

$$\text{The amount of damages the company will pay} = \left(\text{the amount of the damages}\right)\left(\frac{\text{the amount of insurance carried}}{\text{the amount that should be carried}}\right)$$

The amount of insurance that should be carried is equal to:

value of structure • co-insurance percent

SECTION 12-3 Property Tax

1. A property tax is a levy that is charged on property. The amount of the levy is determined by multiplying the assessed value by the rate set by the taxing authority.

   Amount of property tax = (Assessed value of property)(Rate of tax)

2. The assessed value of property is the value of the property for tax purposes. Frequently the assessed value is some fraction of the "market value" of the property. The market value is, in theory, the price that owners could get should they decide to sell it.

   Banks and other lending institutions employ individuals that appraise property. For a house, some of the criteria used to appraise it are the number of square feet in the house, the location, and the type of construction. The appraised value of the house may then be taken as the market value, from which the assessed value can be computed.

3. The tax rate is frequently a sum of rates levied to raise revenue for several needs in the community. Some of the more obvious government financed operations that use the property tax for revenue are:

   a. County government

   b. Schools

   c. County parks and recreation

   d. County law enforcement

   e. County maintained roads and sewers

   f. Payment of county authorized bonds.

   City government may also use the property tax as a means of raising revenue to finance city operations. The taxes paid to city and county governments are deductible from an individual's gross earnings before the federal income tax is computed. That is, property taxes are deductions for federal income taxes.

SECTION 12-4 Income tax

1. The income tax that is collected by the federal government can be levied on any income earned in the United States. The agency responsible for collecting the tax is the Internal Revenue Service, or simply IRS.

   Although states can also levy an income tax, only the federal income tax will be studied in this text.

2. The typical federal income tax form has eight sections. These eight sections, and one item that may be included in each section, are listed below.

	SECTION	ITEM THAT MAY BE INCLUDED IN THE SECTION
a.	*Income*	*Wages earned by the taxpayer*
b.	*Adjustments to income*	*Moving expenses incurred by the taxpayer*
c.	*Itemized deductions*	*Interest expenses*
d.	*Exemption allowance*	*Dependent exemption*
e.	*Regular taxes*	*A tax found with a table*
f.	*Credits against tax*	*Retirement income credit*
g.	*Other taxes*	*Tax on an IRA distribution*
h.	*Refundable credits*	*Income tax withheld*

3. Steps 1 through 9 provide a recommended procedure for completing a personal income tax return.

   Step 1. Find the sum of all sources of income to find a gross income.

   Step 2. Find the sum of all adjustments to income, and subtract the sum from gross income to get adjusted gross income.

Step 3. Find the sum of the itemized deductions subject to limits and subtract the standard deductions to find the net itemized deductions.

Step 4. Subtract the net itemized deductions from the adjusted gross income to get a subtotal.

Step 5. Multiply the number of exemptions by the exemption allowance and subtract the product from the subtotal in Step 4 to get taxable income.

Step 6. Compute the regular tax on taxable income using a tax table or other methods that can be used.

Step 7. Find the sum of credits against tax, and subtract the sum from the regular tax figure obtained in Step 6 to obtain a tax subtotal.

Step 8. Find the sum of other taxes and add the sum to the tax subtotal in Step 7 to get a total tax value.

Step 9. Find the sum of refundable credits, and subtract the sum from the total tax in Step 8 to find an amount of tax due or an amount to be refunded.

The example of a Federal Income Tax Table in Figure 12-4 will be used to find the amount of regular tax on taxable incomes studied in this text. More extensive tables are available from the Internal Revenue Service.

PART B Selected Solutions

1. Step 1. $\frac{75{,}000}{1{,}000} = 75$ — *Face value / 1,000*

   Step 2. Under "Term Life" and on row "Age of Insured" is 19, read \$4.43 — *The insured's age is 19 and the policy is term.*

   Step 3. $75 \cdot \$4.43 = \$332.25$ — *The product is the annual premium.*

   Thus the annual premium is \$332.25.

5. The periodic premium is \$369.90 and the period is quarterly. Thus,

   The rates in Table 12-1 are based on annual premiums. Therefore the quarterly payments are converted to annual.

   $$\text{annual premium} = \$369.90 \cdot 4 = \$1{,}479.60$$

   $\frac{40{,}000}{1{,}000} = 40$ — *Face value / 1,000*

   $$\text{Annual cost per } \$1000 = \frac{1{,}479.60}{40} = \$36.99$$

   Annual premium / Number of \$1,000 in face value

   Using Table 12-1, locate \$36.99 under "Endowment Life" column. Read "Age of Insured" on that row as 36. Thus, the age of the insured was 36 years.

9. The annual premium is \$542.03

   $$\frac{\text{Face value}}{1{,}000} = \frac{67{,}000}{1{,}000} = 67$$

$$\frac{\text{Annual premium}}{\text{Number of \$1,000 in face value}} = \frac{\$542.03}{67} = \$8.09$$

Using Table 12-1, locate $8.09 under "Term Life" column. Read "Age of Insured" on that row as 41. Thus, the age of the insured was 41 years.

13. Step 1. $\frac{55,000}{1,000} = 55$ — *Face value / 1,000*

Step 2. For age 38 and term life, the table factor is 6.76. — *The insured's age is 38 and the policy is term.*

Step 3. Annual premium = $6.76 · 55 = $371.80 — *The product is the annual premium.*

17. Step 1. $\frac{75,000}{1,000} = 75$ — *Face value / 1,000*

Step 2. For age 27 and endowment life, the table factor is 35.81. — *The insured's age is 27 and the policy is endowment.*

Step 3. Annual premium = $35.81 · 75 = $2,685.75 — *The product is the annual premium.*

21. From Table 12-1, for an individual age 45, the table factor for Term Life is 10.55. Thus, it costs George $10.55 per $1,000 insurance. The annual premium is $79.13, therefore

$$\frac{\$79.13}{10.55} = 7.500\text{, to three decimal places}$$

To the nearest dollar, the face value of the policy is $7,500.

25. From Table 12-1, for an individual age 45, the table factor for Endowment Life is 40.38. Thus it costs Wallace $40.38 per $1,000 insurance per year.

Annual premium = $75.71 · 4 = $302.84

$$\frac{\$302.84}{40.38} = 7.500\text{, to three decimal places}$$

To the nearest dollar, the face value of the policy is $7,500.

Exercise Set 12-2 Property Insurance

1. a. Annual premium = $464.00 · 0.98 = $454.72 — *Annual premium rate is 98.00%*

   b. Semiannual premium = $464.00 · 0.5175 = $240.12 — *Semiannual premium rate is 51.75%*

   c. Quarterly premium = $464.00 · 0.2675 = $124.12 — *Quarterly premium rate is 26.75%*

d. Monthly premium = \$464.00 0.0995 *Monthly premium rate is 9.95%*

= \$46.17, to the nearest cent

5. Value of contents = Total - Land - Structure

= \$375,000 - 41,000 - 214,000

= \$120,000

Total premium = premium on structure + premium on contents

= (\$2,140)(0.42) + (\$1200)(0.61)

= \$898.80 + 732

= \$1,630.80

The annual premium on structure and contents is \$1,630.80.

9. Value of contents = Total - Land - Structure

= \$3,700,000 - 1,006,500 - 2,180,000

= \$513,500

Let x = the premium per \$100 of valuation on contents.

Total premium = (Value of structure)(rate on structure)+(Value of contents)(rate on contents)

10,349.91 = (21,800)(0.365) + (5,135)(x)

10,349.91 = 7,957 + 5,135x

2,392.91 = 5,135x

0.466 = x

The premium per \$100 of valuation on contents is \$0.466.

13. Total premium = (\$865.1)(0.59) + (\$187.5)(0.57)

= \$510.409 + 106.875

= \$617.28, to the nearest cent

17. Using the equation:

$$\text{The amount of damages the company will pay} = \left(\text{the amount of the damages}\right)\left(\frac{\text{the amount of insurance carried}}{\text{value of structure} \cdot \text{co-insurance \%}}\right)$$

Let x = the value of the structure.

$$225{,}000 = (225{,}000)\left(\frac{500{,}000}{x \cdot 0.865}\right)$$

$$1 = \frac{500{,}000}{0.865x}$$ *Divide both sides by 225,000.*

$$x = \frac{500{,}000}{0.865}$$ *Multiply both sides by x.*

x = 578,035, to the nearest whole number

The value of the structure is approximately $578,035.

21. Using the co-insurance equation with the following values:

Value of structure = $200,000

Co-insurance percent = 0.80

Amount of insurance carried = $125,000

Amount of loss = $46,500

$$\text{Insurance company will pay} = 46{,}500 \cdot \frac{125{,}000}{200{,}000 \cdot 0.80}$$

$$= 46{,}500 \cdot \frac{125{,}000}{160{,}000}$$

$$= 46{,}500 \cdot 0.78125$$

$$= 36{,}328.125$$

The insurance company will pay $36,328.13 to the nearest cent.

Exercise Set 12-3 Property Tax

1. Assessed value = (Market value)(Rate of assessment)

Assessed value = $36,200 · 0.30

= $10,860

For tax purposes, Edna Coping's home has a value of $10,860.

5. Amount of property tax = (Assessed value of property)(Rate of tax)

Property tax = $64,850 · 0.03079

= $1,996.73 to the nearest cent

The property tax on George Farewell's home is $1,996.73 a year.

9. Adding the individual tax rates:

County	1.1500%	
County parks and rec bonds	0.0025%	
School district	0.0185%	
Sewer district bonds	0.1500%	
County improvement bonds	0.0625%	
Total	1.3835%	*Rate of property tax*

Assessed value = (Market value)(Rate of assessment)

Assessed value = $77,500 · 0.45 *Assessment rate is 45%*

= $34,875

Property tax = (Assessed value)(Rate of property tax)

Property tax = $34,875 · 0.013835 *1.3835% = 0.013835*

= $482.50, to the nearest cent

Exercise Set 12-4 Income Tax

1. Step 1. Gross income $46,926

Step 2. Adjustments to income - 3,390

Adjusted gross income $43,536

Step 3. Itemized deductions - 5,490

Step 4. Subtotal $38,046

Step 5. Exemptions allowance - 4,000

Taxable income $34,046

5. Let x = the amount of the itemized deductions.

Step 1. Gross income $13,269

Step 2. Adjustments to income - 0

Adjusted gross income $13,269

Step 3. Itemized deductions - x

Step 4. Subtotal ($13,269 - x)

Step 5. Exemptions allowance - 5000

Taxable income $ 7,991

Based on Steps 4 and 5 the following equation can be written:

($13,269 - x) - $5000 = $7,991

($13,269 - x) = $12,991

$278 = x

Thus the itemized deductions total $278.

9. Gross income $18,920

Adjustments to income - 510

Adjusted gross income $18,410

Itemized deductions 178

Subtotal $18,232

Exemptions allowance - 3,000

Taxable income $15,232

Thus, Ray's taxable income is $15,232.

13. In the tax table in Figure 12-3, locate the values 26,300 - 26,350 in the column headed "At least - But less than". Since Ray is head of household, in the right column of the group headed "And you are", read the value 4,816.

Thus, Ray's tax is $4,816.

17. Let x = the amount of credits against tax.

Regular tax	$7,008
Credits against tax	- x
	$7,008 - x
Other taxes	+ 288
	$7,296 - x
Refundable credits	-6,018
Tax payable	$1,278 - x

Since it is given that the tax payable is $853, the following equation can be written:

$1,278 - x = $853

$1,278 = x + $853   *Add x to both sides*

$425 = x   *Subtract $853 from both sides*

Thus, the credits against tax was $425.

21.

Regular tax	$7,862
Credits against tax	- 586
	$7,276
Other taxes	+ 1,052
	$8,328
Refundable credits	- 8,211
Tax due	$ 117

Thus, Sally and Bob owe $117 on their taxes.

PART C Chapter Twelve Sample Test

In 1-3, use the information in the following table.

Annual Premium Per $1,000 of Insurance		
Age of Insured	*Term Life*	*Endowment Life*
30	*$4.98*	*$36.05*
31	*5.08*	*36.17*
32	*5.22*	*36.28*

1. Shirley Corey is 31, and this year bought $27,500 in term life insurance. Compute the amount of the annual premium.

1. ______________

2. Sam Nordquist is 30. This year he bought some endowment life insurance. The quarterly payments are $180.25, and there is no additional charge for making quarterly payments. What is the face value of Sam's policy?

2. ______________

3. Geraldine Hammerker, a 32 year old stockbroker, bought term life insurance this year. Her monthly payments are $26.86 and this includes a $2.50 per month service charge. What is the face value of Geraldine's policy?

3. ______________

4. Jerry Hart is 23 and the annual premium on his car insurance is $1173. His insurance company uses the following schedule for payments.

Payment is made	*Percent of annual premium*
Annually	*100%*
Semiannually	*53 1/4 %*
Quarterly	*28 1/2 %*
Monthly	*10 1/3 %*

a. Compute the amount of the semiannual payment. 4. a. ______________

b. Compute the amount of the quarterly payment. b. ______________

c. Compute the amount of the monthly payment. c. ______________

5. Jennifer and Gary Hanover own a home with a value of $53,950, not counting the lot on which the home is built. They value the contents of the home at $34,500. The rates of insurance are $0.58 per hundred dollars of value for the house, and $0.72 per hundred dollars of value for the contents. They want to make quarterly payments, and there is a service charge of $3.00 per payment for quarterly payments.

a. Compute the cost of the annual premium on house and contents. 5. a. ______________

b. Compute the amount of each quarterly payment. b. ______________

In 6-8, use the following information.

In Sierra County the property is assessed at $33\frac{1}{3}$% of market value for tax purposes. The county is currently using the following tax rates.

Fund recipient	*Tax rate*
County general fund	*1.035%*
County bonds	*0.155%*
Sierra Unified School District	*1.405%*
Parks and recreation bonds	*0.015%*
Water and sewage bonds	*0.075%*

6. Sue and Jason Quarles own a farm in Sierra County. Their farm has a fair market value of $349,500.

   a. Compute the assessed value of the farm. 6. a. ________

   b. Compute the tax on the farm for one year. b. ________

7. Home Hardware and Building Supplies in Sierra County has property assessed for taxes at $79,800.

   a. Compute the market value of the property. 7. a. ________

   b. Compute the tax on the property for one year. b. ________

8. Penelope Stanworth last year paid $418.86 property tax on her home in Sierra County.

   a. Compute the assessed value of the home. 8. a. ________

   b. Compute the market value of the home. b. ________

In 9 and 10, supply the missing information:

	Gross Income	*Adjustments to Income*	*Itemized Deductions*	*Exemptions Allowances*	*Taxable Income*
9.	$19,753	$535	$2,946	$5,000	$________
10.	$33,495	$2,250	$________	$4,000	$19,637

In 11 and 12, supply the missing information:

	Regular Tax	*Credits Against Tax*	*Other Taxes*	*Refundable Credits*	*Tax Payable or* < *Refund Due* >
11.	$13,912	$2,150	$760	$13,275	$________
12.	$ 7,469	$583	$126	________	$ 246

CHAPTER THIRTEEN

FINANCIAL STATEMENTS

PART A Summary of Topics

OBJECTIVES

1. *Learn what is meant by an income statement.*
2. *Prepare an income statement.*
3. *Make a vertical percent analysis of an income statement.*
4. *Make a horizontal comparative analysis of an income statement.*
5. *Make a horizontal percent analysis of an income statment.*
6. *Learn what is meant by a balance sheet.*
7. *Calculate equity from a balance sheet.*
8. *Make a horizontal comparative analysis of a balance sheet.*
9. *Compute the current ratio for a balance sheet.*
10. *Compute the quick-test ratio for a balance sheet.*
11. *Compute the inventory turnover ratio.*
12. *Compute the accounts receivable turnover ratio.*
13. *Learn what is meant by break-even analysis.*
14. *Learn what expense values are used to compute break-even.*
15. *Calculate the percent of sales that are variable expenses.*
16. *Calculate the amount of sales necessary for break-even.*
17. *Learn what is meant by inventory.*
18. *Identify the two inventory accounting methods for determining when valuation of an inventory should be taken.*
19. *Identify the two methods ofpricing items in an inventory.*
20. *Learn the four inventory costing methods: specific identification method, average cost method, LIFO, and FIFO.*

SECTION 13-1 The Income Statement

1. An income statement is a schedule that shows (a) the sales and (b) the expenses of a business or operation. An income statement shows whether a net profit or loss for the business or operation occurred during the period covered by the income statement.

2. Equation (1) identifies the result obtained when expenses are subtracted from income.

$$\text{Net profit (or loss)} = \text{Income} - \text{Expenses} \qquad (1)$$

A profit results when Income is greater than Expenses. A loss results when Expenses are greater than Income.

3. To write an expanded version of an income statement, Expenses and Profit are each separated into two categories.

Category 1. Cost-of-goods-sold ⟩ Expenses (3)
Category 2. Operating expenses

Category 3. Gross profit = Income - Cost-of-goods-sold (2)

Category 4. Net profit (or loss) = Gross profit - Operating expenses

Equations (2) and (3) can be called Profit equations. Equation (2) shows that Gross profit is simply the difference between Income and Cost-of-goods-sold. Equation (3) shows that Net profit is the difference between Gross profit and Operating expenses. Operating expenses includes such items as payroll, building and equipment costs, and maintenance.

As will be seen in Section 13-7, the term Cost-of-goods-sold can be abbreviated COGS.

SECTION 13-2 Analyzing the Income Statement

1. A vertical percent analysis is made of an income statement when each item of the statement is expressed as a percent of the total sales for the period of the statement.

   The vertical percent analysis requires that each of the other items in the income statement be divided by the sales figure shown in the statement. The quotient is rounded to two decimal places and written as a percent.

2. A horizontal comparative analysis is made of an income statement by comparing each item of the statement with similar items from a previous income statement or proposed budget.

   The income statement used as the comparative one could be the preceding period, the same period last year, or any other period for which a comparison is desired.

3. To calculate the difference for a horizontal comparative analysis, the subtraction is usually done in the following order:

   Difference = Present income statement - Comparative statement

   If a Comparative statement value is larger than the corresponding Present income statement value, then the difference is negative. To avoid writing negative numbers the symbol <difference> is used.

4. The differences obtained in a horizontal comparative analysis can be used to make a horizontal percent analysis computation. The procedure converts each difference to a percent in which the denominator is the value for that item found on the comparative income statement.

$$\textit{Horizontal comparative analysis percent} = \frac{\textit{Difference in comparative items}}{\textit{Amount from comparative income statement}}$$

The quotient is rounded to two decimal places and then written as a percent.

SECTION 13-3 The Balance Sheet

1. A balance sheet is a schedule that shows the financial position or net worth of a business or individual as of a specified date.

2. A basic balance sheet will have three elements.

Heading 1. Assets - property or cash that is owned.

Heading 2. Liabilities - property or cash that is owed.

Heading 3. Equity - the difference between assets and liabilities.

Equation (4) shows the relationship between Assets, Liabilities, and Equity.

$$\text{Assets} - \text{Liabilities} = \text{Equity} \qquad (4)$$

A balance sheet may separate Asset and Liability items into two or more subcategories. For such balance sheets it is necessary to find the totals of the Asset and Liability items before the Equity equation can be used.

SECTION 13-4 Analyzing the Balance Sheet

1. A horizontal comparative analysis computation can be made of a balance sheet. The procedure is the same as for the Income Statement. That is, the differences are computed for each item in a current balance sheet and the corresponding line items on a comparative balance sheet.

   The choice of which balance sheet to use as the comparative one is a matter decided by the business making the comparison. The equation used to calculate the difference is the same as the one used for the Income Statement.

   $$\text{Difference} = \text{Present balance sheet} - \text{Comparative balance sheet}$$

2. The current ratio is formed by using the current assets and current liabilities from a balance sheet. The current ratio is defined by Equation (5).

   $$\text{Current ratio} = \frac{\text{Current assets}}{\text{Current liabilities}} \qquad (5)$$

   The current ratio is usually written in the form X to 1 or $\frac{X}{1}$, and X is rounded to one decimal place. The ratio can be used to evaluate the ability of an individual or business to pay its debts.

3. The quick-test ratio is formed by dividing the difference between the current assets and inventory by the current liabilities. The quick-test ratio is defined by Equation (6).

   $$\text{Quick-test ratio} = \frac{\text{Current assets} - \text{Inventory}}{\text{Current liabilities}} \qquad (6)$$

   The quick-test ratio is also written in the form X to 1 or $\frac{X}{1}$, and X is rounded to one decimal place. This ratio is also used to evaluate the ability of an individual or business to pay its debts. However, since the ratio is computed without the benefit of inventory, the ratio is more strict, and is sometimes referred to as the acid test ratio. (The phrase probably refers to the practice of using acid to check precious metals).

SECTION 13-5 Further Computations Using the Income Statement and Balance Sheet

1. Turnover describes the cycle Buy goods - Store goods - Sell goods. A new cycle follows the third step in the cycle, namely "Sell goods". Two quantitative measures studied in this section can describe certain aspects of the cycle.

2. The Inventory turnover ratio is calculated by dividing the Cost-of-goods-sold by the Cost of inventory. This ratio is defined by Equation (7).

$$\text{Inventory turnover ratio} = \frac{\text{Cost-of-goods-sold}}{\text{Cost of inventory}} \qquad (7)$$

In Section 13-7, three methods for valuating the cost of inventory are studied. The average cost method is one that is studied at that time.

Notice that the Inventory turnover ratio is also written in the form X to 1, or $\frac{X}{1}$ *, and X is rounded to at most one decimal place.*

3. The Accounts receivable turnover ratio is calculated by dividing the Total sales by the Total accounts receivable. The ratio is defined by Equation (8).

$$\text{Accounts receivable turnover ratio} = \frac{\text{Total sales}}{\text{Total accounts receivable}} \qquad (8)$$

The accounts receivable turnover ratio is usually written in the form X to 1, or $\frac{X}{1}$ *, and X is rounded to at most one decimal place.*

SECTION 13-6 Break-even Analysis

1. Break-even analysis is a procedure to calculate break-even, the amount of sales necessary to pay for the expenses of handling a product (or products) without showing a profit or loss.

   The meaning of the word break-even is consistent with the intuitive notion, namely "no gain, no loss". Notice that Break-even analysis is the procedure to use to compute a break-even situation.

2. Expenses are identified as being of two types:

   Type 1. Fixed expenses. Expenses that usually do not vary, such as building costs (rent or mortgage), salaries of employees on fixed incomes, and utility expenses.

   Type 2. Variable expenses. Expenses that change (increase or decrease) as sales correspondingly change, such as cost-of-goods-sold, commissions, and inventory storage.

   The following three values are needed to calculate break-even:

   1. The total fixed expenses

   2. The total variable expenses

   3. The percent of sales that are variable expenses

3. To write an equation to compute break-even, the following symbols are used:

   S = the amount of sales needed to break-even

   x% = the percent of S that are variable expenses

$$S = \frac{\text{fixed expenses}}{(100 - x)\%}$$

SECTION 13-7 Valuation of Inventory

1. Inventory is the collection of items of personal property that belong to any one of the following categories:

   a. Items that are held for sale in the ordinary course of business (finished products).

b. Items that are in the process of production for sale (work in process).

c. Items used directly, or indirectly, in the production of goods or services to be available for sale (raw materials and supplies).

2. There are basically two inventory accounting methods:

a. The Perpetual Inventory Method

With this method each item in inventory is accounted for. With changes in inventory as a result of a sale out of inventory or a purchase that adds to inventory, an adjustment is instantly made in the inventory and the cost-of-goods-sold.

Many supermarkets and large retail outlets maintain a perpetual inventory method by means of scanners that read codes on the labels of items and a computer makes instantaneous adjustments in the inventory.

b. The Periodic Inventory Method

With this method a "physical inventory" is taken at the end of each period. A period is usually one year, and the inventory is estimated for the interim period by using sales and purchases records.

3. The cost of an item in an inventory is the price paid, or consideration given to acquire it. Cost includes the applicable expenses and changes, directly and indirectly, that were incurred to bring the item into the inventory.

The market price of an item refers to the cost of replacement, by purchase or reproduction. In this text, all figures used will be the cost of the items when they were first placed in the inventory.

4. To compute cost-of-goods-sold (COGS) the following equations will be used:

(a) Beginning Inventory + Purchases = Goods available for sale

(b) Goods available for sale - Ending inventory = Cost-of-goods-sold

5. The four inventory costing methods:

a. Specific identification method

Each item in the inventory is specifically identified by a serial number, or some other way that makes it possible to identify exactly which items remain in inventory. The sum of the costs of these items is the cost of the inventory.

b. Average cost method

Each item in the inventory with the exact same characteristics is assigned the average cost of all the items in the category. This method can be used when the items within a given category are indistinguishable from each other.

c. FIFO (first-in, first-out)

In the FIFO method the assumption is made that the items that have been in stock the longest time are the ones that are sold first. With this method the items within a given category are assumed to be indistinguishable from each other.

d. LIFO (last-in, first-out)

In the LIFO method the assumption is made that the items that have been in stock the shortest time are the ones that are sold first. With this method the items within a given category are assumed to be indistinguishable from each other.

PART B Selected Solutions

Exercise Set 13-1 The Income Statement

1. First find the total expenses

Expenses = \$29,160 + 13,910	*Total expenses*
= \$43,070	*= Cost-of-goods-sold + operating expenses*
With sales = \$44,865,	
Net profit = \$44,865 - 43,070	*Net profit (or loss) = Income - Expenses*
= \$1,795	

Smokey Hills Feed store had a profit of \$1,795 because Income was larger than Expenses.

5. For the given data:

Sales		\$18,426.60
Cost-of-goods-sold	\$11,741.80	
Gross profit		\$ 6,684.80
Operating expenses		5,678.25
Net profit (or loss)		\$ 1,006.55

The placement of these numbers in certain columns is covered in an accounting course. The display shown here is consistent with general accounting practice.

Exercise Set 13-2 Analyzing the Income Statement

1. With sales = \$62,850 and advertising costs = \$3,142, then

$$\frac{\text{Advertising costs}}{\text{sales}} = \frac{3,142}{62,850}$$

$$= 0.049$$

$$= 0.05 \text{ to two decimal places}$$

$$= 5\%$$

Thus, for the time period covered by the income statement, advertising costs were 5% of sales.

5. Using the equation:

Difference = Present statement amount - Comparative statement amount

Difference in sales = $33,030 - 28,974

= $4,056

Since the difference is positive, the present statement sales amount is larger than the sales amount of the comparative statement.

9. Computing the difference,

Difference = $2,100 - 2,000 *Difference = Present amount - Comparative amount*

= $100

Computing the percent,

Percent difference = $\frac{100}{2,000}$ *Percent difference* = $\frac{\textit{Difference}}{\textit{Comparative amount}}$

= 0.05, or 5%

The difference is 5% of the amount shown on the comparative income statement.

Exercise Set 13-3 The Balance Sheet

1. Using Equation (4),

Equity = $85,926.95 - 73,047.68 *Equity = Assets - Liabilities*

= $12,879.27

For the given business, the Equity is $12,879.27.

5. First find the sum of the assets,

Cash	$28,551.90
Inventory	62,143.85
Equipment	126,413.00
Total assets	$217,108.75

Now find the sum of the liabilities,

Note payable	$ 49,126.00
Equipment loans	68,714.00
Total liabilities	$117,840.00

Using Equation (4),

Equity = $217,108.75 - 117,840.00

= $99,268.75 *Equity = Assets - Liabilities*

Exercise Set 13-4 Analyzing the Balance Sheet

1. Using the equation,

   Difference = Present Balance Sheet Amount - Comparative Balance Sheet Amount

   Difference in cash amounts = \$57,867 - 49,228

   = \$8,639

   The current balance shows \$8,639 more cash than the comparative balance sheet. If a given difference is negative, the symbol < difference > will be used.

5. With Current assets = \$654,376 and Current liabilities = \$384,927

   $$\text{Current ratio} = \frac{654{,}376}{384{,}927} \qquad \textit{Current ratio} = \frac{\textit{Current assets}}{\textit{Current liabilities}}$$

   = 1.7, to one decimal place

   The current assets are about 1.7 times the current liabilities.

9. With Current assets = \$327,188, Inventory = \$40,767, and Current liabilities = \$286,421,

   $$\text{Quick-test ratio} = \frac{327{,}188 - 40{,}767}{286{,}421} \qquad \textit{Quick-test ratio} = \frac{\textit{Assets - inventory}}{\textit{liabilities}}$$

   $$= \frac{286{,}421}{286{,}421}$$

   = 1.0, to one decimal place

   The quick-test ratio is 1.0 to 1. Not counting inventory, the assets are the same as the liabilities.

Exercise Set 13-5 Further Computations Using the Income Statement and Balance Sheet

1. With Cost-of-goods-sold = \$43,704, and Inventory = \$10,926,

   $$\text{Inventory turnover ratio} = \frac{43{,}704}{10{,}926} \qquad \textit{Inventory turnover ratio} = \frac{\textit{Cost-of-goods-sold}}{\textit{inventory}}$$

   = 4.0, to one decimal place

   Thus, about \$4.00 in goods are sold for each \$1.00 in inventory.

5. With Sales = \$67,237 and Accounts receivable = \$18,172,

   $$\text{Accounts receivable turnover ratio} = \frac{67{,}237}{18{,}172} \qquad \textit{Accounts receivable turnover ratio} = \frac{\textit{sales}}{\textit{accounts receivable}}$$

   = 3.7, to one decimal place

   Thus, there are about \$3.70 in sales for each \$1.00 in accounts receivable.

9. a. With Average inventory cost = \$70,651 and Cost-of-goods-sold = \$148,369,

Inventory turnover ratio $= \frac{148,369}{70,651}$

$= 2.1$, to one decimal place

Thus, about $2.10 in goods are sold for each $1.00 in inventory.

b. With Total sales = $216,597 and Total accounts receivable = $83,307,

Accounts receivable turnover ratio $= \frac{216,597}{83,307}$

$= 2.6$, to one decimal place

Thus, about $2.60 are made in sales for each $1.00 in accounts receivable.

Exercise Set 13-6 Break-even Analysis

1. a.

Variable expenses	
Wages, production	$13,053
Materials	21,755
Commission	2,176
Total	$36,984

Fixed expenses	
Wages, office	$10,000
Rent	8,500
Property tax	310
Total	$18,810

b. Percent of sales that are variable expenses is x%, and

$x\% = \frac{36,984}{43,510}$ $\qquad x\% = \frac{\textit{Variable expenses}}{\textit{Sales}}$

$= 0.85$, to two decimal places

$= 85\%$

5. Let S = the amount of sales necessary for break-even

x% = the percent of sales that are variable expenses.

With x% = 80% and Fixed expenses = $8,600,

$S = \frac{8,600}{(100 - 80)\%}$ $\qquad S = \frac{\textit{Fixed expenses}}{(100 - x)\%}$

$= \frac{8,600}{0.20}$

$= 43,000$

Thus, for the given data, $43,000 in sales is needed to break-even.

Exercise Set 13-7 Valuation of Inventory

1. Purchases = $66,900

Ending Inventory = $11,350

Cost-of-Goods-Sold = $63,400

Beginning Inventory = x

Beginning Inventory + Purchases - Ending Inventory = Cost-of-Goods-Sold

$$x + 66{,}900 - 11{,}350 = 63{,}400$$

$$x + 55{,}550 = 63{,}400$$

$$x = 7{,}850$$

The cost of Beginning Inventory is $7,850.

5. Beginning Inventory = $55,555

Purchases = $1,885,020

Ending Inventory = $54,828

Cost-of-Goods-Sold = x

Beginning Inventory + Purchases - Ending Inventory = Cost-of-Goods-Sold

$$55{,}555 + 1{,}885{,}020 - 54{,}828 = x$$

$$1{,}885{,}747 = x$$

The Cost-of-Goods-Sold is $1,885,747.

9. a. Beginning Inventory = 427 + 345 + ... + 620

= 4012

Purchases = 85 + 112 + 100 + 267 + 314 + 585 + 585 + 388 + 190 + 190 + 412 + 100 + 100 + 75

= 3503

Cost-of-Goods-Sold = 345 + 516 + 100 + 217 + 620 + 314 + 112 + 188 + 585 + 924 + 85

= 4,006

Let x = the cost of Ending Inventory

Beginning Inventory + Purchases - Ending Inventory = Cost-of-Goods-Sold

$$4{,}012 + 3{,}503 - x = 4{,}006$$

$$7{,}515 - x = 4{,}006$$

$$7{,}515 = x + 4{,}006$$

$$3{,}509 = x$$

The cost of Ending Inventory is $3,509.

b. The Cost-of-Goods-Sold is $4,006.

PART C Chapter Thirteen Sample Test

1. Prepare an expanded income statement using the following information:

Sales	*$212,632*
Cost-of-goods-sold	*164,712*
Sales expenses	*16,218*
Operating expenses	*18,382*

2. The following is an income statement for Hughes Electronics for the month of March.

Sales	*$825,267*
Cost-of-goods-sold	*614,186*
Gross profit	*$211,081*
Operating expenses	*143,802*
Net Profit	*$ 67,279*

To the nearest percent, make a vertical percent analysis of the income statement.

2. a. ____________________

b. ____________________

c. ____________________

d. ____________________

3. Complete a horizontal percent analysis for the following data. Round each percent to the nearest tenth of a percent.

	Actual	*Budgeted*	*Difference*	*% Change*
Sales	*$810,000*	*$850,000*	*$* ________	________
Cost-of-goods-sold	*326,600*	*358,900*	________	________
Gross profit	*$483,400*	*$491,100*	*$* ________	________
Operating expenses	*330,500*	*296,500*	________	________
Net profit	*$152,900*	*$194,600*	*$* ________	________

4. Complete a balance sheet and find the equity for the following data.

Cash on hand	*$825,000*	*Accounts payable*	*$265,000*
Equipment	*300,000*	*Mortgage payable*	*436,000*
Prepaid Ins.	*20,000*		

5. Complete a horizontal comparison analysis on the following balance sheets.

	Present Balance Sheet	*Comparative Balance Sheet*	*Difference Increase/<Decrease>*
Assets			
Cash	*$ 19,500*	*$ 19,700*	________
Accounts rec	*38,000*	*41,000*	________
Inventory	*26,000*	*24,000*	________
Building	*62,800*	*60,000*	________
Equipment	*36,400*	*35,000*	________
Total assets	*$182,700*	*$179,700*	________
Liabilities			
Accounts payable	*$ 17,800*	*$ 18,200*	________
Notes payable	*22,000*	*25,000*	________
Mortgage payable	*44,200*	*45,400*	________
Equity	*98,700*	*91,100*	________
Total liabilities and equity	*$182,700*	*$179,700*	________

6. Compute the following ratios for the balance sheets in Problem 5.

	Cost-of-goods-sold	*Sales*
Present Balance Sheet	*$ 81,500*	*$ 94,200*
Comparative Balance Sheet	*$148,000*	*$152,600*

	Present Balance Sheet	*Comparative Balance Sheet*
a. Current Ratio	________	________
b. Acid-Test Ratio	________	________
c. Inventory Turnover	________	________
d. Accounts-Receivable Turnover	________	________

7. Using Break-even analysis, find the amount of sales necessary for break-even.

Fixed expenses	*Percent of Sales that are Variable Expenses*
$17,800	*80%*

CHAPTER FOURTEEN

BUSINESS STATISTICS

PART A Summary of Topics

OBJECTIVES

1. *Learn the parts of a frequency table.*
2. *Learn how to group a set of data in a frequency table.*
3. *Display data using a bar graph.*
4. *Display data using a line graph.*
5. *Display data using a circle graph.*
6. *Learn how to find the mean of a set of data.*
7. *Learn how to locate the median of a set of data.*
8. *Learn how to identify the mode of a set of data.*

SECTION 14-1 Frequency Tables

1. A frequency table organizes a set of data by grouping the data in classes. Each data point in the set will belong to one and only one class, based on the interval of possible values assigned to each class.

   A frequency table used to organize a given set of data is not necessarily unique. For example, the number of classes into which the data should be put is not rigidly specified. As a general rule, the larger the number of data points, the more classes will be used. However, rarely is less than 5 classes or more than 15 classes used in any table.

   The interval of the classes in a table is also not rigidly fixed. By interval we mean the range of possible values that can be assigned to a class. To illustrate, if the data points with values 5, 6, 7, 8 and 9 are assigned to Class 1, then the interval of this class is 5. In general, all classes in a table should have the same interval.

2. The frequency tables in this text have six headings:

   a. Class number - the number assigned to each class in the table.

   b. Class limits - the smallest and largest possible values that can be put in a given class.

   c. Class boundary - a value halfway between an upper limit of one class and a lower limit of the adjoining class.

      The class interval can be calculated by computing the difference in the boundaries of the class.

   d. Class mark - the middle value in a class.

   e. Tally marks - a mark that is recorded for each data point in the class.

   f. Class frequency - the number of data points in a class.

   A frequency table may include other headings, such as Relative Frequency and Cumulative Frequency. For this text only the six headings listed will be studied.

SECTION 14-2 Graphs

Many different types of graphs for sets of data can be studied in a course in statistics. For this text only the Bar graph, Line graph, and Circle graph are studied.

1. Three of the principal parts of a bar graph are:

   a. A Horizontal Axis. On the horizontal axis the possible values of one of the two variables in the data set are listed.

      Remember, horizontal is similar to horizon. Therefore the horizontal axis is the bottom axis that runs from left to right.

   b. A Vertical Axis. On the vertical axis the possible values of the second variable in the data set are listed.

      Usually the smaller values are listed on the lower end of the vertical axis and the larger values on the upper end.

   c. The Bars. The center and height of each bar indicate the values on the horizontal and vertical axes, respectively, that are paired with that bar.

2. The following five steps provide a recommended procedure for constructing a bar graph.

   Step 1. Draw horizontal and vertical line segments that are perpendicular to each other and join at one end.

   Step 2. Label each axis to identify which variable is to be listed on it.

   Step 3. Locate evenly spaced points on the horizontal axis, one for each data point to be listed on the axis, and label each point.

   Step 4. Locate evenly spaced points on the vertical axis and label each point.

   Step 5. Draw a bar, centered on each point on the horizontal axis, to a height corresponding to the value of the second variable, using the scale on the vertical axis.

3. A line graph is frequently used to show a trend in the values of one variable with respect to time. For example, line graphs can be used to show how business sales rise, or fall, over some interval of time. Such line graphs can be used to make predictions as to future values of the plotted variable.

4. The following six steps provide a recommended procedure for constructing a line graph.

   Step 1. Draw a horizontal and vertical axis for the graph.

      This is similar to Step 1 for constructing a bar graph.

   Step 2. Label each axis. If one of the variables is time, label the horizontal axis time.

   Step 3. Locate evenly spaced points on the horizontal axis, one for each data point to be listed on the axis. If this variable is time, the points will be labeled with units of time, such as days, months, or years.

   Step 4. Locate evenly spaced points on the vertical axis, and label each point with the possible value of the second variable.

   Step 5. Place a dot above each labeled point on the horizontal axis to a height that corresponds with the paired value.

      The height of the dot above the horizontal axis shows the value of the second variable paired with the indicated value of the variable on the horizontal axis.

Step 6. Connect the dots in succession with line segments.

The up (and down) directions of the line segments show the corresponding rise (and fall) of the values of the variable on the vertical axis.

5. A circle graph can be used to show how two or more variable quantities make-up a whole. Such a graph can be used to show what percents of some quantity belong to two or more categories.

   A circle graph is frequently used to show where the tax dollar for the federal government comes from. That is, what percent is from personal income tax, what percent is from tax on corporations, and so on. Similarly, a circle graph is used to show where the federal tax dollar is spent. That is, what percent is for military spending, what percent for interest on the national debt, and so on.

6. The following three steps provide a recommended procedure for constructing a circle graph.

   Step 1. With a compass or other tool draw a circle with a convenient radius.

   Step 2. Multiply each percent by 360° and round each product to the nearest degree.

   Step 3. With a protractor, draw angles in the circle with measures equal to the values obtained in Step 2.

   The relative sizes of the sections of each "piece of the pie" reflect the percents of the whole in the categories identified. That is, if a Category A owns 30% of the whole, then in the circle, the piece of the pie that represents Category A is 30% of the area of the circle.

Section 14-3 Measures of Central Tendency

1. For a set of quantitative data, the arithmetic mean is $\bar{x}$ (read "x-bar"), and

$$\bar{x} = \frac{\text{the sum of the data points}}{\text{the number of data points}}$$

   The term arithmetic mean is used to distinguish it from a geometric mean. However, in statistics the term mean is used, and it is understood that the arithmetic mean is intended.

2. The median of a set of quantitative data is the value of the item of data in the middle of the ordered array of data points. The symbol $\tilde{x}$ (read "tilde x") will be used for a median.

   The following two steps can be used to find the median for a set of data.

   Step 1. Array the data from smallest to largest, or from largest to smallest.

   Step 2. Locate the middle number in the array.

   If the number of data points is odd, then the median is the data point in the center of the array.

   If the number of data points is N, then $\frac{N+1}{2}$ is the median in the array. To illustrate, if the array has 13 data points, then $\frac{13+1}{2} = \frac{14}{2} = 7$ is the number of the data point that is the median.

   If the number of data points is even, then the median is the average of the two data points in the center of the array.

If the number of data points is N, then $\frac{N}{2}$ *and* $\frac{N+2}{2}$ *are the two data points in the middle of the array, and the average of these numbers is the median. To illustrate, if the array has 24 data points, then* $\frac{24}{2} = 12$ *and* $\frac{24+2}{2} = 13$ *are the two middle data points. The average of these two data points is the median.*

3. The <u>mode</u> of a set of quantitative data is the data point that has the greatest frequency. If two different data points have the greatest frequency, then they are both called modes, and the data is said to be bimodal.

PART B Selected Solutions

Exercise Set 14-1 Frequency Tables

1. Five classes are specified, and a class interval of 3 grams. Thus, three values are possible for each class. The lower limit for Class 1 is 240 grams, therefore 240, 241 and 242 are the possible values for Class 1; 243, 244 and 245 are the possible values for Class 2; and so on. The boundaries for Class 1 are 239.5 and 242.5. Notice that 242.5 - 239.5 = 3.0, the specified class interval. The class mark for Class 1 is $\frac{240 + 242}{2} = 241$.

Class No.	Class Limits	Class Boundaries	Class Mark	Tally Marks	Class Freq.
1	240 - 242	239.5 - 242.5	241	11	2
2	243 - 245	242.5 - 245.5	244	1111	4
3	246 - 248	245.5 - 248.5	247	~~1111~~ 1	6
4	249 - 251	248.5 - 251.5	250	~~1111~~ ~~1111~~ 11	12
5	252 - 254	251.5 - 254.5	253	~~1111~~ 1	6

<u>Note 1.</u> The class interval 3 is the difference between succeeding lower class limits. That is,

243 - 240 = 246 - 243 = 249 - 246 = 252 - 249 = 3

Therefore, adding 3 to the lower limit of any class will yield the lower limit of the next class. The same is true for the upper limits.

<u>Note 2.</u> The class interval 3 is the difference between succeeding lower class boundaries. That is,

242.5 - 239.5 = 245.5 - 242.5 = 248.5 - 245.5 = 3.0.

Therefore, adding 3 to the lower boundary of any class will yield the lower boundary of the next class. The same is true for the upper boundaries.

<u>Note 3.</u> The class interval 3 is the difference between succeeding class marks. That is,

244 - 241 = 247 - 244 = 250 - 247 = 253 - 250 = 3.

Therefore, adding 3 to the class mark of any class will yield the class mark of the next class.

5. Five classes are specified, and a class interval of 10. The lower limit of Class 1 is 60, therefore 60, 61, 62, ..., 69 are the possible values for Class 1; 70, 71, 72, ..., 79 are the possible values for Class 2; and so on. The boundaries for Class 1 are 59.5 and 69.5, and the class mark for Class 1 is $\frac{60 + 69}{2} = 64.5$. Use the class interval 10 to generate the class limits, class boundaries, and class marks of the other 4 classes.

Class No.	Class Limits	Class Boundaries	Class Mark	Tally Marks	Class Freq.
1	60 - 69	59.5 - 69.5	64.5	111	3
2	70 - 79	69.5 - 79.5	74.5	~~1111~~ 111	8
3	80 - 89	79.5 - 89.5	84.5	~~1111~~ ~~1111~~ ~~1111~~	15
4	90 - 99	89.5 - 99.5	94.5	~~1111~~ ~~1111~~ 11	12
5	100 - 109	99.5 - 109.5	104.5	11	2

Exercise Set 14-2 Graphs

1. Label the horizontal axis Grades, and locate five evenly spaced points on the axis. Label these points A, B, C, D and F from left to right. Label the vertical axis Number, or Number of Students, or simply Frequency. Locate six evenly spaced points on the axis. Since the maximum number to be shown on this axis is 110, it is sufficient to label these points 20, 40, 60, 80, 100 and 120 from bottom to top. Now draw bars centered on A, B, C, D and F to heights that reflect the given values. The result is the following bar graph.

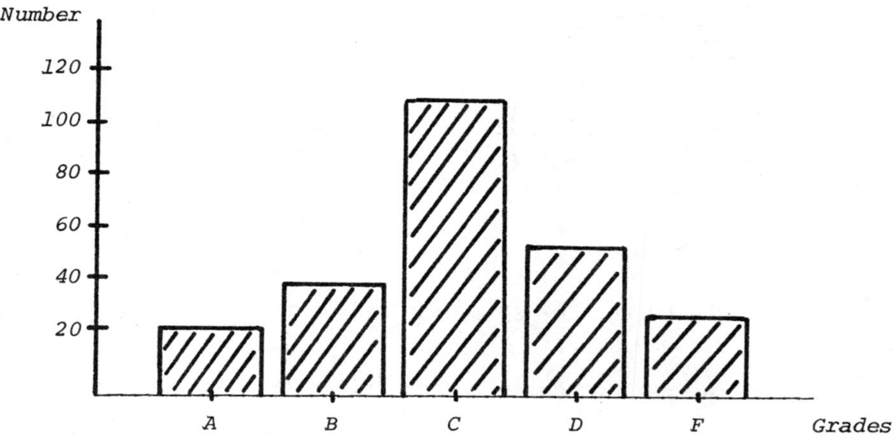

5. Label the horizontal axis Month, and locate twelve evenly spaced points on the axis. Label these points J (for January), F(for February), and so on up to D (for December) from left to right. Label the vertical axis Number, or Number of Customers. The smallest value to be plotted is 468 and the largest value to be plotted is 562. It is therefore sufficient to locate about twelve equally spaced points on this axis, and label these points 460, 470, 480, and so on up to 570. Now locate points above J, F, M, ..., D that show the number of customers for each month. Connect successive dots with line segments to obtain the following line graph.

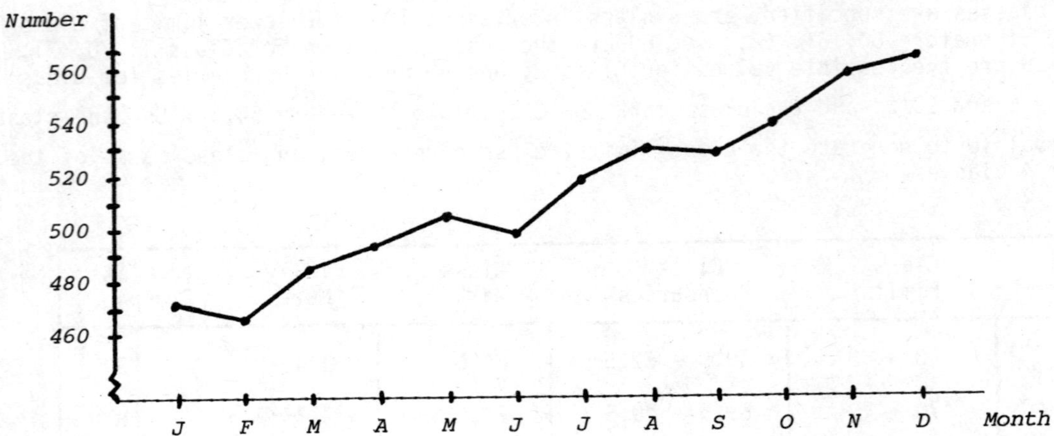

7. Draw a circle with a fairly large radius. Now multiply each percent by 360° to determine the size of the angle needed to represent each of the five grade distributions.

For 15%: $0.15(360^{\circ}) = 45^{\circ}$

For 22%: $0.22(360^{\circ}) = 79^{\circ}$, to the nearest degree

For 40%: $0.40(360^{\circ}) = 144^{\circ}$

For 20%: $0.20(360^{\circ}) = 72^{\circ}$

For 3%: $0.30(360^{\circ}) = 11^{\circ}$, to the nearest degree

Draw angles with the indicated degrees in the circle, and label the sectors accordingly A-grades, B-grades, and so on.

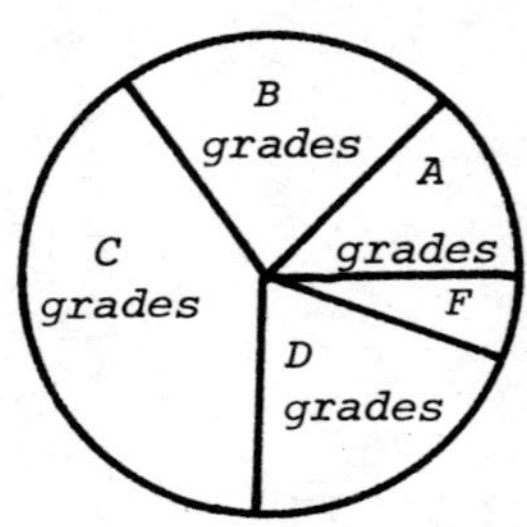

9. a. As shown in Figure 14-B, the percent of softwoods milled that is Douglas fir is 31.7%.

b. As shown in Figure 14-B, the percent of softwoods milled that is White fir is 8.3%.

c. The percent of the total that is Ponderosa Pine is 12.2%.

12.2% of 26,812

= 0.122(26,812)

= 3,271, to the nearest whole number.

About 3,271 million board feet of Ponderosa Pine was harvested.

Exercise Set 14-3 Measures of Central Tendency

1. a. $\bar{x} = \frac{23.70 + 15.25 + 31.60 + \ldots + 28.80}{8}$

$= \frac{215.1}{8}$

$= 26.89$, to two decimal places

The mean for the set of data is \$26.89.

b. Arraying the data from smallest to largest:

\$15.25, \$16.90, \$18.40, \$23.70, \$28.80, \$29.10, \$31.60, \$51.35

The two data points in the middle are \$23.70 and \$28.80.

$\tilde{x} = \frac{23.70 + 28.80}{2}$

$= 26.25$

The median for the set of data is \$26.25.

3. a. $\bar{x} = \frac{12 + 10 + 10 + 15 + \ldots + 12}{20}$

$= \frac{224}{20}$

$= 11.2$

The mean score on the quiz was 11.2.

b. Arraying the data from smallest to largest:

6, 8, 9, 9, 9, 10, 10, 10, 10, 10, 12, 12, 12, 12, 13, 13, 14, 15, 15, 15

The median is the average of the two data points in the middle, namely 10 and 12.

$\tilde{x} = \frac{10 + 12}{2}$

$= \frac{22}{2}$

$= 11$

The median score on the quiz was 11.

c. The score 10 has the greatest frequency, namely 5. Thus the mode score was 10.

9. a. $\bar{x} = \frac{89 + 84 + 82 + 81 + \ldots + 55}{14}$

$= \frac{1028}{14}$

$= 73.4$, to one decimal place

The mean number of games won was 73.4.

b. Arraying the data from smallest to largest:

55, 64, 64, 65, 68, 69, 70, 71, 79, 81, 82, 84, 87, 89

The median is the average of the two data points in the middle, namely 70 and 71.

$$\tilde{x} = \frac{70 + 71}{2}$$

$$= \frac{141}{2}$$

$$= 70.5$$

The median number of games won was 70.5.

c. The data point 64 has the greatest frequency, namely 2. Thus the mode is 64.

PART C Chapter Fourteen Sample Test

1. The following weights were obtained from a sample of 40 cans of pears taken from a large production run at Hunts Cannery in Geyserville. The weights are in grams per can.

404	409	403	410	406
407	406	404	408	407
405	401	413	402	410
408	410	410	412	410
410	400	412	413	409
412	413	409	404	412
413	400	406	411	410
413	410	411	409	413

Put this data in a frequency table of five classes, a lower limit of 400 for Class 1 and a Class interval of 3 grams.

Class No.	*Class Limits*	*Class Boundaries*	*Class Mark*	*Tally Marks*	*Class Freq.*
1					
2					
3					
4					
5					

2. A stock broker has a reputation for recommending common stock with high dividend rates. The following data represent the dividend rates of some stocks recommended last month.

3.9	5.8	6.8	7.0	2.6	4.4	3.4	3.6
4.8	4.8	4.8	7.0	5.1	4.0	3.7	5.0
5.6	5.6	2.7	5.6	3.9	5.4	2.8	7.3

For this set of data:

a. Compute $\bar{x}$ to one decimal place. 2. a. ____________

b. Find $\tilde{x}$. b. ____________

c. If one exists, state the mode. c. ____________

3. Forty tulip bulbs were selected from a shipment of bulbs from Holland, Michigan and planted. The flowers produced were as follows:

a. 15 were red flowers (R)

b. 10 were blue flowers (B)

c. 6 were yellow flowers (Y)

d. 6 were white flowers (W)

e. 3 were varigated flowers (V)

Construct a bar graph for this set of data using the axis below.

4. At a large university the student body was polled on the basis of their political party affiliation. The following percents were obtained:

a. 40% were registered Democrats

b. 30% were registered Republicans

c. 20% were registered Independents

d. 10% were registered with some other party

Construct a circle graph that shows the composition of political party affiliation at the university polled.

CHAPTER FIFTEEN

AN OVERVIEW OF COMPUTERS IN BUSINESS

PART A Summary of Topics

OBJECTIVES

1. *Learn what is a spreadsheet program.*
2. *Study some uses of spreadsheet programs.*
3. *Learn what is a data base management program.*
4. *Study some uses of data base management programs.*
5. *Learn what is meant by a single-application program.*
6. *Study some examples of single-application programs.*
7. *Learn what is a word processing program.*
8. *Study some of the advantages of a word processor.*
9. *Learn what is meant by a graphics program.*
10. *Observe some of the types of graphs that can be generated by a graphics program.*

In this chapter some of the ways in which computers can be used in business are surveyed. The discussion can only touch on the more important details related to each of the applications. A more indepth study of any one of these applications can only be covered in a course dedicated to that goal.

1. A <u>spreadsheet program</u> is a computer program that will allow the computer to manipulate large quantities of numerical data and display them in a format similar to that of accounting worksheets.

   The significant feature of a spreadsheet is that it is an array of rows and columns that separate the display into <u>cells</u>, or <u>boxes</u>. By labeling the rows and columns, an individual cell can be <u>specifically identified</u> by the column and row that contains that cell.

   A cell in a spreadsheet can contain a number, word, or formula. The word or number can be displayed in the cell, but the formula will not be displayed.

2. The following list identifies some of the features of a spreadsheet program that make it a very useful tool for many business situations:

   1. A spreadsheet program can replace the need for paper, pen, and calculator, since the computer records all data and instantly performs all calculations.

   2. A spreadsheet program unerringly remembers all the relationships among entries in the worksheet and rapidly performs all calculations with one command.

   3. A spreadsheet program overcomes the limitations of paper worksheets, in that many programs offer hundreds of columns and thousands of rows.

   4. A spreadsheet program permits the user to change one or more entries in the worksheet and then instantly observe the effect of such a change on the other entries in the worksheet.

   5. Two or more spreadsheets can be connected to provide information between the worksheets.

3. The following five steps can be used to make a spreadsheet:

Step 1. Determine a format for the spreadsheet by answering questions a through c.

a. How many columns will be needed?

b. How many rows will be needed?

c. How many spaces will be used for each column?

Step 2. Enter all column and row headings into the spreadsheet.

Step 3. Enter all the known values.

Step 4. Enter the formulas for the amounts to be calculated.

Step 5. Calculate the results of the formulas.

4. A data base management program is a computer program that can search a data base for items within the file with a defined characteristic. Other terms used for such a program are Data-File Management Program and Information Retrieval System. The information contained in such a program is fundamentally non-numerical.

Some of the information in the program may be numbers, such as addresses and phone numbers. But these numbers are basically descriptive and are not used in formulas to arrive at some quantitative result, such as is done in spreadsheets.

5. A single-applications program is a computer program that is designed to perform one specific task. Three examples of such programs are a through c.

a. Payroll - Computes gross and net pay, maintains all payroll records, and prints the checks.

b. Inventory Control - Maintains records of increases and decreases in inventory of merchandise and material.

c. Depreciation - Computes depreciation schedules for property using any one of the several methods of depreciation.

6. A word processing program stores information that is entered in the form of language text. The program permits the text to be reviewed, reorganized, and modified and provides a printout.

Some of the features of a word processor are identified in a through d.

a. Words, sentences, or paragraphs can be deleted, inserted, or rearranged.

b. Words or phrases that have been used repeatedly throughout the text can be changed or removed with a single instruction.

c. The text can be checked for correct spelling of every word in the text.

d. The program provides several options for the format of the printed copy.

Some of the options that can be adjusted in the format are left and right margins, top margin, line spacing, number of lines per page, paragraph indentation, pitch, and automatic page numbering.

7. A graphics program enables a person to use the speed of the computer to generate some type of graph for a given set of data.

Some of the types of graphs that can be constructed with a graphics program are:

a. Vertical bar graph - the bars extend up and down.

b. *Horizontal bar graph - the bars extend right and left.*

c. *Broken-line graph - the series of line segments show possible trends in the data.*

d. *Pie chart - the areas of the various sectors show how two or more categories share the whole of something.*

These graphs were constructed manually in Chapter 14. Using a graphics program makes the work much simpler.

PART B Selected Solutions

1. A spreadsheet is a display that has rows and columns.

   The rows and columns make the cells that are the significant feature of a spreadsheet.

5. A spreadsheet program can organize and store numerical data.

   One of the major differences between the spreadsheet program and data base management program is that the data stored in a spreadsheet is numerical, and in a data base it is non-numerical information.

9. A data base management program basically stores non-numerical information.

   The word basically is used since numbers can be stored in a data base. However, the numbers are used for identification purposes (such as telephone numbers and addresses), and not for computation.

13. A spreadsheet program permits the user to change several entries in a worksheet and then instantly observe the effect of the changes on other entries in the worksheet.

    This use of a spreadsheet program can be called the "What if" phenomenon. That is, a single change can be made in one of the entries, and a single command to the computer will correspondingly change the other entries in the spreadsheet.

17. Computers can be linked to each other by means of telephone lines or other communications media.

    As a consequence, it is possible to link computers on opposite sides of the world.

21. The success of the computer in business is based in part on b. its ability to process information quickly and accurately.

    It should be noted, however, that the cost of computers is annually getting lower, so that the cost is also becoming an attractive feature.

25. Which of the following can be entered into any cell of a spreadsheet?

    a. words, b. formulas, and c. numbers.

29. Which computer program would be used to store the tax information on real estate property in a county?

    c. A data base management program.

    The data stored is primarily information. Specifically, information such as location, address, owner, appraised value, and so forth.

33. The column headings are

A: End of Year

B: Depreciable Base

C: Depreciation Expense

D: Accumulated Depreciation

E: Book Value

The row headings are 1 through 5 for the five years of depreciation.

A completed spreadsheet is given below.

	A	B	C	D	E
1	*End of*	*Depreciable*	*Depreciation*	*Accumulated*	*Book*
2	*Year*	*Base*	*Expense*	*Depreciation*	*Value*
3					
4	*1*	*$9,000*	*$3,000* *(B4)•(5/15)*	*$3,000* *(B4)•(5/15)*	*$6,000* *(B4)-(C4)*
5	*2*	*$9,000*	*$2,400* *(B5)•(4/15)*	*$5,400* *(D4)+(C5)*	*$3,600* *(B5)-(D5)*
6	*3*	*$9,000*	*$1,800* *(B6)•(3/15)*	*$7,200* *(D5)+(C6)*	*$1,800* *(B6)-(D6)*
7	*4*	*$9,000*	*$1,200* *(B7)•(2/15)*	*$8,400* *(D6)+(C7)*	*$600* *(B7)-(D7)*
8	*5*	*$9,000*	*$600* *(B8)•(1/15)*	*$9,000* *(D7)+(C8)*	*Ø* *(B8)-(D8)*

PART C Chapter Fifteen Sample Test

1. Prepare a depreciation schedule using a spreadsheet program and the double-declining balance method for the following item:

Item Name: *Ice cream maker*
Cost New: *$9,750*
Scrap Value: *$ 750*
Useful Life: *5 years*

	A	B	C	D	E
1					
2					
3					
4					
5					
6					
7					
8					

2. a. Identify what a word processing program can do.

b. List some of the features of a word processing program that makes it an asset to many business operations.

3. Identify and discuss three single-application programs that can be utilized by many businesses.

ANSWERS TO SAMPLE TESTS

Chapter One

1. 5,072,620
2. Twenty-five million, three thousand, nine hundred sixty
3. 647,110,000
4. 647,000,000
5. 21,076
6. 29,080
7. 2,808
8. 123,525
9. 13,855
10. 15,540,000
11. 426
12. 7/6
13. 25/24
14. 3/7
15. 3/5
16. 27
17. 1/3
18. 47/12 or 3 11/12
19. 55/36 or 1 19/36
20. 1/16
21. 32 ounces
22. $40.50
23. 10 1/2 inches
24. 24 gallons
25. $81 1/2

Chapter Two

1. 8.75
2. 0.0136
3. Twelve and nine hundredths
4. Nine hundred fifty-three hundred-thousandths
5. 38.198
6. 1.8751
7. 86.4
8. 18.922
9. 7.235
10. 14.592
11. 3.98
12. 0.17
13. 5000
14. 308.67
15. 18/5
16. 29/4
17. 1.85
18. 0.8125
19. 2.364
20. 0.214
21. 2.72
22. 2.7189
23. 0.583
24. 0.76

Chapter Three

1. Yes
2. No
3. Yes
4. 6
5. 10
6. 9
7. 3
8. 15
9. A = 38
10. F = 158
11. M = 4.5
12. $\frac{9}{60} = \frac{3}{20}$
13. $\frac{6}{100} = \frac{3}{50}$
14. $\frac{10}{60} = \frac{1}{6}$
15. $\frac{\text{2 gallons}}{\text{625 sq. ft}}$
16. $\frac{\text{3 hours}}{\$20.25}$
17. x = 5
18. y = 3/4
19. 3.5 hours
20. $144.00

Chapter Four

1. 45%	2. 230%	3. 18.9%
4. 3.5%	5. 0.109	6. 1.75
7. 0.12	8. 0.009	9. 1/5
10. 1/12	11. 8/5	12. 3/40
13. 0.375	14. 37.5%	15. 3/20
16. 15%	17. 9/40	18. 0.225
19. 2.2	20. 220%	21. 112.5
22. 7626	23. 0.93	24. 5.08
25. 2400	26. 25%	27. 1200
28. 8%	29. 30%	30. $340.00

Chapter Five

1. $308.88	2. $326.83	3. $424.00
4. $379.71	5. $1,071.68	6. $467.74
7. $418.98	8. $372.50	9. $287.70
10. $28.66	11. $116.48	12. $63.97
13. $157.75	14. $122.40	15. $188.05
16. a. $22.38 b. $ 2.56 c. $ 8.63	17. a. $57.17 b. $ 6.53 c. $22.05	18. a. $101.50 b. $ 1.52 c. $ 5.13

Chapter Six

1. a. 2% b. 20 days	2. a. 5% b. 20 days	3. a. 3 1/2 % b. 20 days
4. a. 2% b. 40 days	5. a. 4 1/2 % b. 10 days	6. November 10
7. November 1	8. November 20	9. November 16
10. November 11	11. a. $20.25 b. $654.75	12. a. $282 b. $6768
13. a. $2710 b. $132,790	14. $191.25	15. $6,111.97
16. $10,077.60	17. a. $70,616 b. $72,800	18. a. $840 b. $126 c. $714
19. a. $768 b. $1632	20. a. $45 b. $165	

Chapter Seven

1. a. $15.00
   b. 25%

2. a. $72.70
   b. 50%

3. a. $706
   b. 33 1/3 %

4. a. $7.45
   b. $44.70

5. a. $37.74
   b. $185.74

6. a. $1,080
   b. $4,320

7. a. $1.59
   b. 20%

8. a. $29.68
   b. 22.4%

9. a. $387.50
   b. 8 1/3% or 8.3%

10. $12.00

11. $326.25

12. $4,240

13. a. $2.25
    b. $20.25

14. a. $90.00
    b. $285.00

15. a. $1,124.70
    b. $13,871.30

16. $6.00
    50%
    33.3%

17. $20.00
    25%
    20%

18. $182.00
    $ 28.00
    15.4%

19. a. $8.80
    b. $12.32

20. $26.86

21. 25%

22. $77.45

23. a. $0.87
    b. $22.67

24. a. $7.29
    b. $139.89

25. a. $337.20
    b. $5,957.20

Chapter Eight

1. a. 219
   b. 215

2. a. 51
   b. 50

3. a. 286
   b. 283

4. a. 74
   b. 73

5. a. 180
   b. 179

6. $645.75

7. $896.40

8. $7,200

9. 14.6%

10. 2 1/2 years

11. $8.25

12. $10.31

13. $115.17

14. $21.94

15. $265.60

16. $241.50

17. 10.4%

18. a. $798.90
    b. $810.00

19. 2.5 years

20. a. $6,372

    b. $29,628

Chapter Nine

1. $2562.50
   $2626.56
   $2692.23
   $2759.53

2. a. $7,153.84
   b. $2,153.84

3. a. $18,952.68
   b. $6,452.68

4. a. $30,674.97
   b. $6,924.97

5. $11,806.48

6. $3,403.28

7. $183.33
   $251.52
   $181.03
   $253.82

8. $117.36

9. $36.42

Chapter Nine Cont'd

10. a. $12,730.93
    b. $5,530.93

11. $24,586.56

12. a. $469.33
    b. $115,479

Chapter Ten

1. a. 325; 1; 325
   b. 205; 2; 205
   c. 111; 2; 111
   d. 248; 1; 248
   e. 59 ; 2; 59

2. a. 26%
   b. 12%
   c. 9%
   d. 18%
   e. 8%

3. a. $ 0
   b. 0
   c. 0
   d. 14
   e. 23
   f. 20
   g. 16
   h. 16

4.

Ending balance on bank statement	$262.93
Add the deposit made but not shown on bank statement	+ 713.72
	$976.65
Subtract the total of the outstanding checks	- 288.50
	$688.15
Enter the check book balance	- 692.65
	- 4.50

The $4.50 discrepancy is the result of the service charge not being deducted on the check register. To reconcile the two balances, deduct $4.50 on the check register, and the corrected balance is $688.15

5. a. $197.00
   b. $492.50
   c. $689.50
   d. $591.00

6. a. $83.44
   b. $60.11
   c. $48.44

7. a. $1636.28
   b. $12.96

8. a. $375
   b. $12,875

Chapter Eleven

1. a. $3.70
   b. 11.8

2. a. 35%
   b. 21%

3. $2700

4. $4,246.24

5. a. $581.63
   b. $154,080

6. $412.50

7. $518

8. 15.2%

9. $8,976

10. a. $2,100
    b. $4,200
    c. $3,515

11. $4,350

Chapter Twelve

1. $139.70

2. $20,000

3. $56,000

4. a. $624.62
   b. $334.31
   c. $121.21

5. a. $561.31
   b. $143.33

6. a. $116,500
   b. $3,128.03

7. a. $239,400
   b. $6,427.89

8. a. $15,600
   b. $46,800

9. $11,272

10. $7,608

11. $<753>

12. $6,766

Chapter Thirteen

1.

Sales		*$212,632*
Cost-of-goods-sold	*164,712*	
Gross profit	*$ 47,920*	*$ 47,920*
Sales expenses	*$ 16,218*	
Operating expenses	*18,382*	
Total expenses	*$ 34,600*	*$ 34,600*
Net profit (or loss)		*$ 13,320*

2. a. $\frac{\text{Cost-of-goods-sold}}{\text{sales}} = 74\%$

   b. $\frac{\text{Gross profit}}{\text{sales}} = 26\%$

   c. $\frac{\text{Operating expenses}}{\text{sales}} = 17\%$

   d. $\frac{\text{Net profit}}{\text{sales}} = 8\%$

3.

Difference	% Change
< 40,000 >	*< 4.7% >*
< 32,300 >	*< 9.0% >*
< 7,700 >	*< 1.6% >*
3,400	*11.5%*
< 41,700 >	*< 21.4% >*

4.

Assets		
Current assets		
Cash	*$825,000*	
Total		*$825,000*
Fixed assets		
Equipment	*$300,000*	
Total		*$300,000*
Other assets		
Prepaid insurance	*$20,000*	
Total		*$ 20,000*
Total assets		*$1,145,000*
Liabilities and Equity		
Current liabilities		
Accounts payable	*$265,000*	
Total		*$265,000*
Long-term liabilities		
Mortgage payable	*$436,000*	
Total		*$436,000*
Equity		*$444,000*
Total liabilities and equity		*$1,145,000*

5. Difference
Increase/<Decrease>

$ <200>
<3,000>
2,000
2,800
1,400
3,000

$ <400>
<3,000>
<1,200>
7,600
3,000

6.

	Present Balance Sheet	Comparative Balance Sheet
a.	2.2 to 1	2.0 to 1
b.	1.9 to 1	1.8 to 1
c.	3.1 to 1	6.2 to 1
d.	2.5 to 1	3.7 to 1

7. $89,000

Chapter Fourteen

1.

Class No.	Class Limits	Class Boundaries	Class Mark	Tally Marks	Class Freq.
1	400-402	399.5-402.5	401	1111	4
2	403-405	402.5-405.5	404	~~1111~~	5
3	406-408	405.5-408.5	407	~~1111~~ 11	7
4	409-411	408.5-411.5	410	~~1111~~ ~~1111~~ 1111	14
5	412-414	411.5-414.5	413	~~1111~~ ~~1111~~	10

2. a. 4.8

b. 4.8

c. 4.8 and 5.6 (Bimodal)

3.

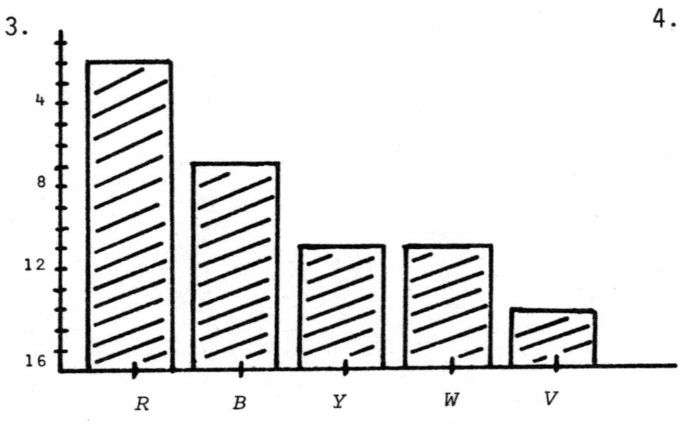

4.

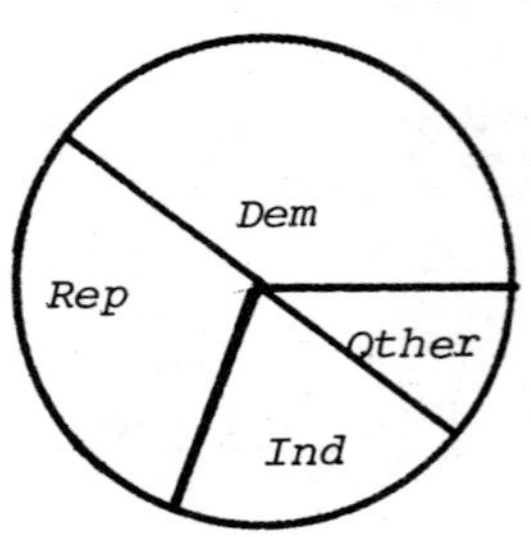

Chapter Fifteen

1.

	A	B	C	D	E
1	*End Of*	*Depreciable*	*Depreciation*	*Accumulated*	
2	*Year*	*Base*	*Expense*	*Depreciation*	*Book Value*
3					
4	*1*	*$9,750.00*	*$3,900.00* *(B4) · 2/5*	*$3,900.00* *(B4) · 2/5*	*$5,850.00* *(B4) - (C4)*
5	*2*	*$5,850.00* *(B4) - (C4)*	*$2,340.00* *(B5) · 2/5*	*$6,240.00* *(D4) + (C5)*	*$3,510.00* *(B5) - (C5)*
6	*3*	*$3,510.00* *(B5) - C5)*	*$1,404.00* *(B6) · 2/5*	*$7,644.00* *(D5) + (C6)*	*$2,106.00* *(B6) - (C6)*
7	*4*	*$2,106.00* *(B6) - (C6)*	*$ 842.40* *(B7) · 2/5*	*$8,486.40* *(B6) + (C7)*	*$1,263.60* *(B7) - (C7)*
8	*5*	*$1,263.60* *(B7) - (C7)*	*$ 505.44* *(B8) · 2/5*	*$8,991.84* *(D7) + (C8)*	*$ 758.16* *(B8) - (C8)*

2. a. A word processing program stores information that is entered in the form of language text.

   b. 1. With a word processor words, sentences, or paragraphs can be deleted, inserted, or rearranged. Such changes can be made and the design of the text will automatically be corrected for each change.

      2. With a word processor the text can be checked for correct spelling. (Technical words that may be unique to a given text can be added to the computer dictionary, and the spelling of these words can be checked throughout the text.)

      3. Words, or phrases, that have been used repeatedly throughout the text can be changed or removed with a single instruction.

      4. Several choices for format can be provided with a word processing program.

3. a. A payroll program. Such a program computes gross and net pay for every employee. The program also maintains the employee records with such information as cumulative earnings, totals of FICA, FIT and SIT deductions.

   b. An inventory program. Such a program keeps a perpetual inventory record of items added to, or removed from a given inventory.

   c. A depreciation program. Such a program keeps a record of the depreciable base, the current years depreciation, the accumulated depreciation, and any other item of information relevant to the depreciation of each item of depreciable property in the business.